Essential Water

Essence of Life and Health

A Proven Guide to Understand the Impact of Water on Your Health and How to Choose the Best Type of Water for You and Your Family

Micah Bright

Publisher M Teodora Quiroga

ISBN 978-1-917060-06-6 (paperback)

ISBN 978-1-917060-07-3 (eBook)

ISBN 978-1-917060-08-0 (Hardcover)

Grab Your Bonus today!

Take charge of your health with our **checklist** for securing great water quality. Ensure better hydration, improved well-being, and a positive impact on the planet—in a few simple steps.

Your journey to better water and better YOU,

starts here!

https://www.micahbrightauthor.com/bonus

Table of Contents

Introduction

What Is Water?

We believe we know water. We absorb it from practically everything we eat or drink. We use it for swimming, washing, and floating. We hope it will start, or wish it would stop falling on our heads. We are aware that, depending on the heat and perspiration, we must drink several liters of it every day; otherwise, we would be driven and then tormented by thirst. We are aware that everything that leaves our body, including breath, tears, and spit, contains water. We are aware that we have to bring our pets and cattle to water each day and that we must water our plants in the event that rain fails. We are aware that most things that are moist and plump are alive, while most things that are dry and shriveled are either dead or will live again after being exposed to water. We are aware that rivers have a downward flow, contain fish, and are inhabited by otters and platypuses. We are aware of the ocean's size, strength, and the presence of sharks and whales. Since we sprinkle water on the ground or dab it on our hair during rituals, we are aware of the power and symbolism of water.

Without actually thinking about it, we are rather knowledgeable about water. We are intuitively aware of its vital role in physiology (perspiration and thirst), ecology (fish and farming), magic (rituals and dreams), economics (we frequently have to purchase it), and power (our susceptibility to people in positions of control). However, this common experience only represents a small portion of the reality of water and our relationship to it, as well as our shared effort to comprehend and utilize it in all its forms.

Water has been on Earth for a very long time, and the evolution of the planet has always depended on its unique characteristics. However, how did water become available in such enormous and fortuitous amounts, and where did it come from? Broadly speaking, water should be common because its molecules are composed of two of the most common elements of ordinary matter in the universe: oxygen, which is the third most common element (after helium, at 23%), and hydrogen, which makes up 75% of all elements.

Moments after the universe first began, hydrogen emerged from the primordial chaos of the Big Bang. However, oxygen and all the other elements with larger and heavier atoms originated from complex thermonuclear events deep within stars way later in time. All the elements within these stars were dispersed far into space when they eventually perished, with the larger ones exploding as supernovas like SN 2006gy, the brightest ever observed. There, they gathered via the gradual pull of gravity into successive star generations, whirling clouds that would eventually give rise to planets and all the other solid entities in the universe, including comets and asteroids.

It so happened that some of these clouds held more water than others and that certain clouds had the ideal conditions to hold onto their water after they solidified. A big planet's strong gravitational field would hold it in place, and the water wouldn't split back into individual hydrogen and oxygen atoms if there wasn't enough radiation from neighboring stars. Water would have eventually disappeared if there had been enough radiation because the lighter hydrogen atoms would have generally drifted off into space, leaving oxygen to mix with other elements. Water would remain on the solid planet as well at low enough temperatures since rock and deep-frozen water are immobile until they warm up.

It appears plausible that at least some of Earth's water appeared during the early stages of the planet's formation when comets attacked it. However, there is evidence for an alternate mechanism as well; tiny comets composed only of water are still bombarding Earth's atmosphere. Hundreds of what appear to be very-high-altitude holes or vapor trails can be seen every day in ultraviolet satellite photographs of the Earth's atmosphere. A group led by Louis Frank, a physics professor at the University of Iowa and a world expert on violent charged particles, plasmas, and auroral imaging of the Earth and other planets in the solar system, has been examining these photographs since the mid-1980s. Frank and his colleagues interpreted the photos as depicting the fragmentation of tiny comets. They imagined these as loosely packed, 20–40 ton "snowballs" that, upon approaching Earth, disintegrate due to fast electrostatic erosion before vaporizing in the high atmosphere (Caldecott, 2008).

Whichever cosmic mysteries and fortune made the Earth the blue planet, it is teeming with life because of it. Our planet is also said to exist in the "Goldilocks" zone, which has many implications. Earth is far and near enough to the sun that water can exist in all three phases on the earth's surface. Entrepreneur Elon Musk is on a mission to terraform Mars, but it's not easy. While there is water on Mars, it is problematic when it comes to supporting life.

This life-giving substance that we take for granted is something of a wonder. In order for this water to make its way into our homes, it has to go through various infrastructure and processes. There is also the fact that most of the Earth's surface is covered in saltwater, and humans are freshwater creatures. Therefore, it takes a lot of engineering to make water accessible and convenient. This means that our water can pick up contaminants along the way and become compromised.

This book is a guide on what measures we can take to make sure that our drinking water is not only safe to drink but also improves our health. You will come across different types of water that you probably haven't heard of before and be informed about the different systems and equipment that you can buy to treat your water. You will also be provided with additional resources. You will be informed and entertained, but mostly, you will walk away with a new appreciation for this wonderful substance.

Bob McCauley (2006) writes in The Miraculous Properties of Ionized Water:

> The person who is hydrated . . .
>
> is always healthier . . .
>
> is less toxic . . .
>
> is more energetic . . .
>
> functions more efficiently overall. . .
>
> retains less water in the extremities (legs, feet, hands) . . .
>
> is more mentally alert. . .
>
> is less prone to disease . . .
>
> is better able to fight disease of every kind. . .

Let's find out why.

Chapter 1:

Understanding the Significance of Water for Our Health

A young seminary student in Dillingen, Bavaria, was near death in the fall of 1849. The young man, who was once strong, had lost weight. Something rattled him: a cough. His sheets were saturated with sweat every night. He frequently discovered blood on his pillow in the morning.

Sebastian Kneipp was the student; he was the impoverished son of a weaver from a little Bavarian hamlet. He suffered greatly after moving from his hamlet to the cramped, unhygienic surroundings of the seminary and hostel. The previously fit young man developed tuberculosis due to lack of exercise, a fresh diet, and fresh air. Throughout the nineteenth century, tuberculosis was widespread and incurable.

Sebastian Kneipp frantically looked through the seminary library for treatment after his doctors had given up on him and he was on the verge of death. He happened to come across an old pamphlet that suggested a cold-water remedy. With little to lose, he took the booklet's recommendation and dove into the frigid Danube River three times a week. He went to bed without taking off his clothes because he didn't have a towel. To the amazement of professors, doctors, and other pupils, this inhumane treatment worked on him and healed him.

Later on, he created a comprehensive health system that he named the *Five Pillars of Health.* This took Europe and the world by storm. At the time, *The Washington Post* listed Sebastian Kneipp as one of the most influential individuals in the world (Fleckenstein & Weisman, 2007).

We've come a long way since the days of consumption. Modern-day medical interventions have made tuberculosis an easily treatable disease, with no icy-cold Danube swimming necessary. Kneipp's story shows us the importance of water for our overall health. In this chapter, we dive into the importance of water, what happens when we don't get enough of it, and how important it is for detoxification.

The Importance of Hydration

Oceans gave rise to all life, and the sea is still present in our cells today. Water makes up over 70% of our bodies, by no means a minor component in maintaining overall health. Water, within and outside our bodies, heals and rejuvenates us because of our oceanic past. The salinity of our cells now is precisely the same (0.9%) as it was the first time a membrane with nucleic acids developed around some saltwater to form a cell.

After emerging from the oceans over four million years ago, our ancestors eventually gave rise to upright humans as we know them today. However, we can never venture far from water.

What, then, can be accomplished by one basic oxygen atom and two hydrogen atoms that make up the water molecule H_20? Hydrogen and oxygen atoms are arranged polarly in water molecules; hydrogen has a positive electrical charge, and oxygen has a negative one. This means that the water molecule is able to draw in a wide variety of other molecules, including salt (NaCl). Because hydrogen attracts negative chloride ions and negative oxygen attracts positive sodium ions, water can dissolve salt. Hydrophilic substances are those that dissolve readily, such as sugar and salt, whereas hydrophobic substances are those that don't like oil. All this is to say that water is an excellent solvent that facilitates the movement of dissolved materials, particularly minerals, throughout the body, which is why it is so beneficial. Furthermore, the electrical and magnetic properties of water are vital to the operation of cells. We can live for a few weeks without food but only a few days without water. We can only be blessed with regeneration, inside and outside, by water. Not only can it improve our mood, but it can also enhance our appearance. Every cell in the body needs water for critical metabolic and hormonal processes.

Within your body, water fulfills numerous vital functions (Fleckenstein & Weisman, 2007): maintaining the body's pH (acidity/alkalinity) balance; lubricating joints and cushioning intervertebral discs; enabling chemical and enzymatic reactions within the body; regulating body temperature and many other functions, thereby promoting balance, harmony, and health; transporting molecules from the outside to inside your body (digestion), thereby nourishing your body;

eliminating toxins from your body; and making blood circulation possible.

The word "hydration" has been used intentionally in the title of this sub-section. Hydration is not just about drinking water. It means drinking enough of it to ensure that our bodies function optimally.

You must consume the recommended amount of water each day in order to bathe your billions of cells in this beautiful fluid. Your urine will seem colorless if you drink too much water because it dilutes important body components, including salts, minerals, and electrolytes. Your urine will seem dark because your kidneys will be unable to flush away waste items if you drink too little water. When is enough, *enough*? For an individual of average height and weight, seven cups of water per day is a general estimate of the appropriate amount. You'll need more when exercising or when the weather is hot. When you get the perfect balance, your urine should be pale yellow.

The impact of adequate hydration is astounding. You'll be able to focus more clearly and get rid of the mental fog that so many of us experience. Because every system in your body will perform more efficiently at the cellular level, you will have better digestion, elimination, and toxin clearance. You'll get deeper, better sleep. I'm not kidding. Allow water to accomplish what no medication or herbal treatment can. When your fascial system, muscles, and joints are well lubricated, your flexibility increases. Unbelievably, approximately 31% of our bones are made of water. You'll get rid of edema and bloating. No ankle edema! Your clothes will fit more comfortably. Water makes your skin look younger and more beautiful because it adds moisture to the skin. As we optimize our waste systems, inflammation decreases.

You are aware that water eventually winds up in your stomach. There, some of that water is passed through your digestive system, and the rest is transferred to your bloodstream to nourish your tissues. It so happens that there are different pathways by which water gets to your organs and other parts of your body besides your bloodstream. Fascia, which are layers of spongy tissue that surround your muscles, organs, blood vessels, nerves, and bones, do more than just wrap and keep your entire body together. Additionally, it has a sophisticated water distribution system that delivers water exactly where it is needed.

On a larger scale, water aids in the body's ability to fend off certain chronic illnesses. In fact, water is by far the most crucial nutrient to take into account when examining the causes of chronic problems. It unquestionably contributes to a number of health problems that we will all face as we become older.

The Effects of Dehydration

I drink a lot of water, you could be thinking. *How is it even possible that I'm dehydrated?*

Even if we believe we drink enough water, the numerous moisture-stealing aspects of modern life leave us feeling mildly exhausted. If a healthy body were compared to a plant, which is rooted in nutrient-rich soil, absorbs minerals and moisture, absorbs sunlight and water, and grows abundantly, then today's dehydrated bodies would be like withered leaves and dry, shriveled stalks.

Humans are said to lose two to three liters (or sixty to one hundred ounces) of water every day as a result of respiration, perspiration, bowel movements, and urine. Keep in mind the proverb that states, "What goes in must come out." To keep our bodies in a state of homeostasis, or equilibrium, the delicate balance of water lost from the body and water taken in must be maintained. If you aren't replenishing the lost water, your brain releases hormones to redirect water from non-life-sustaining places to maintain the proper function of more vital organs, such as your heart, liver, and brain. It is quite possible to suffer from low-grade dehydration without realizing it because thirst is not always an early warning sign of the condition. Dehydration can occur without your realization.

Consider an average workday at the office when you hardly get out of your chair. Your body finds it difficult to flush out waste materials and get water into your cells while you are sedentary for extended periods of time. Because sitting reduces the amount of water in our bodies, we are physically becoming dehydrated.

Dehydration can have a significant effect even at very tiny levels; cognitive decline is detectable at levels as low as two percent. That is a water loss of less than one liter or roughly thirty-three ounces. Just enough to make you lose some of your senses. A sense of appreciation is low, and anxiety decreases. Life has lost its vibrancy. Those are effects when your brain is only two percent dry. And most of us experience that at some point in our lives. We frequently experience low-grade dehydration by three o'clock and near-emptiness by nine o'clock. We dry out repeatedly, daily, throughout our lives. Aging is accelerated daily by this (Cohen & Bria, 2018).

You can get sick from dehydration in practically every way possible. Consider joint pain, migraines, and postsurgical pain. Recent studies conducted by eminent universities, top medical facilities, and even the U.S. Army demonstrate that even a little dehydration can exacerbate these types of symptoms. It also makes it difficult for you to concentrate and makes you feel hungrier (Cohen & Bria, 2018).

The best indicator of dehydration isn't always thirst. Often, when we're feeling tired or have a headache coming on, we should be drinking water instead of thinking about what to eat. According to research, these negative symptoms are likely your brain's internal "alarm system," alerting you to the need for immediate water. Your odds of being dehydrated are high if you're experiencing headaches, irritability, lack of focus, or other unpleasant symptoms (Cohen & Bria, 2018). Furthermore, the issues go beyond feeling dizzy and having a dry mouth. According to research, dehydration is implicated in many, if not most, illnesses. Dehydration is directly linked to a wide range of symptoms. These complaints include: migraines and other types of headaches; fatigue and weakness, both general and associated with illnesses such as fibromyalgia; irritable bowel syndrome; constipation; insomnia; heart disease; type 2 diabetes; acid reflux; and dementia, especially Alzheimer's disease.

Cardiovascular Disease

Would you believe that your blood vessels can be affected by slight dehydration, as little as a two percent decrease in hydration levels, in the same manner that smoking cigarettes does? According to a study from the University of Arkansas, young men who don't drink enough water quickly see a reduction in their endothelium's (the lining of their blood vessels) capacity to dilate and contract, two processes that are critical for normal blood flow (Cohen & Bria, 2018). Imagine the consequences for older individuals or those who have heart disease and risk factors like diabetes. It should come as no surprise that even mild dehydration increases the risk of having a heart attack.

Your blood thickens when its water content decreases. Healthy blood should have the consistency of wine rather than ketchup, according to cardiologist Stephen Sinatra, MD (Cohen & Bria, 2018). But if you're not drinking enough water, that's just how thick your blood can get, and very rapidly. Thick blood increases the effort required by the heart to pump blood, which can weaken the heart muscle and raise blood pressure. Additionally, it compels your body to expend more energy—energy that could be better spent in other parts of your body, including your brain—to get blood through your arteries and capillaries. The end consequence is a confluence of heart and blood vessel problems, together with cellular inflammation, which paves the way for cardiovascular disease.

Diabetes

You're already more likely to become dehydrated if you have type 1 or type 2 diabetes. This is due to the fact that elevated blood sugar levels, or hyperglycemia, can result from your body's inability to produce insulin (as in type 1 and advanced type 2 diabetes) or from your body's insensitivity to insulin (type 2). Why is that a problem? Your kidneys will attempt to eliminate elevated blood sugar from your system by producing more urine. You guessed it: dehydration is the result of over urinating.

It's a vicious cycle: When your blood is thicker and less hydrated, your blood sugar levels are likely to rise since there isn't enough water in your system to dilute all that sugar. This can cause you to become even more dehydrated, which can cause more issues with blood sugar. When your body starts burning fat for energy and releases large amounts of potentially poisonous blood acids called ketones, it can lead to a dangerous condition known as ketoacidosis. This disease is caused by excessively high blood sugar and low insulin levels.

When you are chronically dehydrated, the little water you drink goes directly to your brain, which is the most important organ for your survival. As a result, your blood's water content decreases, which raises blood sugar levels. Your health won't likely be harmed if this just occurs sometimes. However, chronically elevated blood sugar levels brought on by dehydration harm your body's delicate cells. They become less sensitive to the effects of insulin as a result. That's insulin resistance, and type 2 diabetes can develop really quickly.

Digestion

Increasing your water consumption and eating more meals high in water will help ease your digestive troubles, regardless of whether you have cramps or constipation on a regular basis. Possibly the main cause: Water and meals heavy in water, like veggies and chia, also help soften stool and your food during digestion. These foods are also strong in fiber, which aids in digestion. Additionally, soft stool helps you get rid of waste without straining or using laxatives since it moves through your system more quickly and easily, reducing bloating and stomach pain. Since laxatives also dehydrate, they counteract this effect by retaining water, which makes bloating worse.

According to Dr. Batmanghelidj's (2005) book *Your Body's Many Cries for Water*, water can also help with more severe digestive problems like acid reflux or ulcers. Batmanghelidj had spent the latter part of the 1970s as a political prisoner in Iran. Using nothing but water and electrolytes in the form of salt or sugar, he treated and cured many of his fellow inmates during that period, many of whom had peptic ulcer disease. Adequate water consumption is essential if you suffer from a chronic digestive issue such as Crohn's disease, colitis, or irritable bowel syndrome. These disorders induce diarrhea, which causes your body to lose more water than usual.

Deep Vein Thrombosis

When a blood clot forms deep in your veins, usually in one of your legs, you may get Deep Vein Thrombosis (DVT). Travelers are most commonly affected, but other factors that increase your risk include being pregnant, taking estrogen-based birth control, being overweight or obese, being elderly, having poor circulation in your lower legs, or having recently undergone surgery. DVT is a dangerous illness that can result in pulmonary embolisms, which are frequently fatal blocks of blood vessels in the lungs. Two ways that drinking water can prevent DVT are by preventing blood clots and thickening and by making you feel the urge to urinate more frequently. That's why it's crucial to drink lots of water instead of alcohol-containing drinks that dehydrate you when you're traveling by automobile or flying. This also applies if any of the previously listed DVT risk factors apply to you.

Pulmonary Disorders

Water maintains the lubrication and proper operation of all the internal organs of your body, including your lungs and airways. Water thins the mucous membrane lining your throat and lungs. A proper amount of mucus aids in the body's processing and removal of toxins, pollutants, and other chemicals that may be harmful to the lungs, both healthy lungs and those with conditions like Chronic Obstructive Pulmonary Disorder (COPD). Thick, unhealthy mucus filters fewer chemicals that hinder breathing, which is why dehydration exacerbates COPD (as chronic bronchitis or emphysema).

Cognition

Many of us are driven to act by the need to *feel* better. And there are many ways that water can help you with that. Drinking water can help you feel clearer and more focused nearly instantly. People with mild dehydration showed decreased concentration levels and did poorly on tests measuring focus and cognition. On the other hand, when they were completely hydrated, they did well on these tests. People's moods were also affected by dehydration.

There is a connection between the onset of Alzheimer's disease and prolonged dehydration. We now know that diabetes and Alzheimer's share pathophysiology, and a number of medical professionals are now referring to Alzheimer's as "type 3 diabetes." Insulin resistance, inflammation, oxidative stress, obesity, and metabolic syndrome are a few of the causative factors. We think that staying hydrated is the first line of defense against all these threats.

The first line of treatment for preventing numerous illnesses, including concussions (also known as mild traumatic brain injury, or TBI for short), is adequate and appropriate hydration. Adequate hydration is especially significant for children and in sports. The consistency of your brain is like gelatin. Cerebrospinal fluid within your skull protects it from regular knocks and bumps. Your brain may slam against the inside walls of your skull in a forceful blow to your head, neck, or upper body. A sudden collision or fall, for example, can also induce abrupt acceleration or deceleration, which might result in brain damage. More water equals greater protection and cushioning.

Our ERs are overrun by cases of mild TBI. ER visits for concussions resulting from sports-related injuries have sharply increased. Athletes' pre-existing decreased brain resilience as a result of widespread environmental pollutants and dietary deficiencies may be connected to the rise in sports-related concussions. These illnesses represent the body's incapacity to reestablish balance and result in an "increased susceptibility to what would previously be considered innocuous concussions." In these new circumstances, staying hydrated is essential for both defense and water's primary purpose of preserving homeostasis.

Chronic Pain

At least one in five people have chronic pain, and many more get it occasionally (Cohen & Bria, 2018). Water consumption can be a preventative measure as well as a treatment for dehydration, which is known to be the cause of some pains, such as migraines and cramping muscles. However, hydration can still be beneficial in more complicated pain situations like acute injuries, joint aches, and menstrual cramps. Studies reveal that dehydration exacerbates the discomfort you're feeling. It is logical.

Water is diverted when you're dehydrated: It goes directly to your heart, brain, and other essential organs that require water to survive, as opposed to your tissues and joints. This can lead to waste products like lactic acid building up, stiffening of the tissues and joints, and inflammation that can cause discomfort. Furthermore, proper hydration reduces pain levels by calming down brain activity connected to pain, while dehydration increases it.

Water's Role in the Detoxification of the Body

Not getting enough rest or deep sleep? Join the club. In any given week, almost half of all Americans don't get enough sleep. However, taking a tablet isn't a permanent answer, and using sleep aids has its risks. Instead of taking medication, drink water. There are many fundamental benefits of water with regards to sleep. First off, it prevents snoring by keeping your mouth and nasal passages wet. Even if you don't remember it the next day, leg cramps can wake you up. Avoid them by being properly hydrated.

Water consumption during the day is crucial for improving sleep quality at night. The majority of our body's detoxification process happens when we're asleep. And the key component of that detoxifying process is water. Your body moves more important bodily fluids—such as interstitial and cerebrospinal fluid—through your lymphatic system, brain, spine, and other important regions when you sleep. This increase in fluid flow aids in the removal of toxins and certain metabolites from your body and brain, including beta-amyloid protein, a factor in Alzheimer's disease. The procedure functions better the more hydrated you are. Thankfully, you don't need to down two liters of water before bed to sleep better. You're already taking adequate steps to support your body's detoxification and promotion of restorative, deep sleep if you choose hydrated foods and drink water throughout the day.

Sleep and beauty go hand in hand; you've probably heard the expression "Get your beauty rest" a million times. This aphorism has a new meaning, and that meaning is detoxification. Scientists were taken aback when they discovered a drainage system concealed in specific brain cells. Though it only removes waste from the brain, this system functions similarly to the lymphatic system. It is referred to as the glymphatic system. Similar to the nocturnal sanitation team, it increases flow by 60% when people are sleeping. It seems as though our minds utilize this ideal moment to clear out clutter after being freed from "thought traffic." Knowing exactly when and how the brain clears waste and activates the lymphatic system is a critical first step in efforts to make it work more efficiently. Experts are working on this. Sufficient quantities of hydration are necessary for all of this circulation and flow. Who knew that part of hydration included reclining, shutting our eyes, and nodding off?

Water does not just detoxify our bodies through sleep; it also does so through sweat. Exercise and sports ought to be a regular part of your day. Some people exercise somewhat more than others, and some people sweat slightly more than others. Some sweat much more. Gym addicts also need to pay more attention to their water. This is why it's so fantastic. While we don't give sweat the respect it deserves, it is an extremely important and intricate outflow system that works with numerous other bodily networks. Sweating is an energy-intensive and complicated process. Your kidneys alone have the capacity to burn around four hundred calories per day, depending on your body's resting metabolic rate (i.e., how many calories you burn while at rest). A large portion of this energy is used to maintain the proper balance of fluid intake between your body's cells and organs. Your body's natural thermostat relies on sweating to help you cool off quickly and prevent overheating of your tissues, blood, and organs.

One of the numerous ways your body naturally detoxifies is by sweating, which aids in the removal of some substances from your bloodstream. Additionally, it serves as a vehicle for your body's natural pheromones, or "sex hormones," that have a scent and affect who you attract and are drawn to. The dermis, or layer of skin just beneath the portion of your skin that is touchable contains glands that create sweat. Although sweat glands are found throughout your body, they are most concentrated in your forehead, armpits, palms, and soles of your feet.

Sweat's precise makeup varies from day to day and from person to person. However, 99 percent of it is water. What makes up the remaining 1%? In addition to ammonia, sugar, and trace amounts of minerals like calcium, potassium, magnesium, iron, zinc, copper, and some water-soluble vitamins, the electrolytes include sodium and chloride. Not only must we replace water lost through perspiration, but we also need to replace electrolytes. This is ideally addressed by adding some natural sea salt to our water bottles.

Water can also help detoxify our bodies from the outside in. Kneipp, with whom we began this chapter, stumbled upon this discovery. While this book is focused on drinking water, I would be remiss not to mention that immersing our bodies in water can detoxify us.

Immersion in cold water enhances the body's natural healing process. Beyond eliminating sweat, grime, dead skin cells, bacteria, and viruses, cold water has further benefits: Boost defenses against cancer and infections; increase your metabolism to combat type 2 diabetes, obesity, gout, rheumatic disorders, depression, and more; boost your mood and motivation; give your glands (thyroid, adrenals, ovaries/testes) a boost; normalize blood pressure, reduce chronic pain, strengthen tired, irritated nerves, train and enhance blood circulation, detoxify your body, combat lethargy, repair and tone your skin, and deepen your breathing; help with sleeplessness; enhance kidney function; diminish edema and swelling; boost the immune system through improved lymphatic circulation; reduce stress by controlling your autonomic nervous system; control body temperature, preventing excessive perspiration and chronically cold hands and feet; maintain healthy hair; improve hemorrhoids and varicose veins; and lessen aches and pains (Fleckenstein & Weisman, 2007).

In addition to its diuretic effects, taking a saltwater warm bath detoxifies the body, improves circulation and blood flow to the skin, eases joint and muscular aches, and calms and sedates dry, itchy skin. In actuality, it treats or resolves a variety of persistent skin issues as well.

Water has the ability to smooth stones and create canyons. So, consider the potential benefits it could have on your health if you drink the highest quality water. Water has power that should never be undervalued. Your health might be significantly improved by simply taking the time to understand the different qualities of water and how they might affect you. In the next chapter, we explore these differences.

Chapter 2:

Different Types of Water and Their Impact on the Body

In the winter of 1512, Juan Ponce de Leon had everything. Twenty years earlier, at the age of seventeen, he had embarked on Christopher Columbus's second trip to the New World as a deckhand. Following his return home, Ponce de León made the decision to remain and pursue his dreams of riches. Ponce de León was a hard-charging, ambitious, and furious combatant, as his biographer subsequently put it: "a man spirited, sagacious and diligent in all warlike matters." This would guarantee his quick development in the burgeoning Spanish empire, where incredible wealth was only waiting to be stolen. In 1509, he was named governor after spearheading the conquest of Puerto Rico and seizing the island for Spain. He was now formally established, bearing both riches and domains.

But life at the top was not fulfilling. A conquistador's existence did not consist of administering lands for the Crown and doing official tasks. He wanted more. He set out once more in pursuit of new experiences, but this time, he was looking for much more than just land and gold. The Indians in the area had told him tales of an amazing place. This place promised to fulfill the poet's deepest desires and ambitions. They told him of a land far to the north that was rich in wealth and full of pleasures, but most of all, it had a river that was so pure that anyone who dipped in it would become young again.

This was the alchemist's dream! All one needed to do was locate this blessed place and bask in the abundance of unending wealth and eternal youth! Indeed, some of the prehistoric Indians said that it was unnecessary to travel so far in search of these restorative waters because a fountain with the same incredible and incalculable properties could be found on the island of Bimini, which is part of the Bahamas group, and is located far out in the ocean. With warm credulity, Juan Ponce de León listened to these stories. He was becoming older, and the average lifespan didn't seem long enough for his grand ambitions. If he were to plunge into this amazing fountain or gifted river, what would he not be able to accomplish in the additional years? His body, battered and worn from war, would be restored to its youthful strength, freshness, and suppleness, and his head would still hold the wisdom and knowledge of age.

Captivated, Ponce de León asked King Ferdinand II of Spain for a charter that would allow him to benefit from any new lands he might find. After the monarch granted his request, Ponce de León sailed in March of 1513 with three ships and sixty-five soldiers. He traveled the Caribbean in search of these enchanted waters. On Easter Sunday, he came upon a new place that he named *Pascua de Florida*, or the "flowers of Easter." He also discovered the Gulf Stream, which was crucial to later navigators. He was the one who discovered the American continent, not Christopher Columbus. His Florida finding paved the path for later European immigration to North America. His career should have been a huge success by most accounts.

But Ponce de León's main mission was unsuccessful. He sustained injuries during an Indian attack in Florida during his second voyage. In 1521, he passed away in Havana, Cuba, from his wounds. The daring adventurer, who made significant contributions to mapping the Caribbean's waters and terrain, never saw the *Fountain of Youth* that he so desperately sought. Ironically, Ponce de León's hunt has become immortal despite his inability to partake in the fabled elixir. Among all adventure stories, his search for the fabled *Fountain of Youth* has become one of the most well-known (Salzman, 2017).

The story above is only a legend. However, it is a legend that permeates all societies throughout history, which makes it important. From ancient Mesopotamia and Alexander the Great, to the Norse god Odin and Christianity, there has always been a strong link between drinking specific waters and healing or wisdom.

As James Salzman (2017) puts it in *Drinking Water: A History*:

There are few universals in the human condition, but sacred wells may be one of the rare constants. They have been found in virtually every culture around the world and in virtually every era. From earliest times, natural water sources have been linked with mystical healing and powers.

Natural water is only one of the different waters that we are going to discuss in this chapter. We'll be discussing tap, mineral, alkaline, electrolyzed, artesian, deuterium-depleted, and rainwater. Then, we offer a comprehensive comparison of all these different types of water and highlight the best ones.

Tap Water vs. Bottled Water

Water that has been packed in airtight containers for human use is known as bottled water. It can originate from municipal tap water supplies, icebergs, springs, or aquifers. Conversely, tap water is described by the US Environmental Protection Agency (EPA) as water obtained from a public water system, which is a system for supplying water via pipes or other built conveyances.

When deciding whether to use tap or bottled water, keep in mind that the main distinction between the two is how they are delivered—not the source or quality, which might differ greatly for either. Although tap water is almost always less expensive and, depending on the situation, more environmentally friendly, neither is intrinsically safer or healthier.

Most disparities between tap and bottled water in terms of safety, health, cost, regulation, and environmental friendliness can be attributed to their packaging and delivery and the particular conditions under which it manifests for various brands and tap water systems. So, it's not as easy as one might believe to decide between tap and bottled water.

What is Bottled Water?

"Water that is intended for human consumption and sealed in bottles or other containers with no added ingredients, except that it may contain a safe and suitable antimicrobial agent as well as limited levels of fluoride," is how the US Food and Drug Administration (FDA) defines bottled water. "Water intended for human consumption put into bottles or containers intended for sale" is how the EU Drinking Water Directive defines bottled water (Bottled water vs. tap water, n.d.). Almost any type of packing, such as bottles, cans, or cartons, can be referred to as a container.

The US FDA states that "bottled water" must originate from a reliable source. Based on their geological origins and the methods used for treatment, the FDA defines well water as any water that can be drawn out of a borehole or drilled into the ground that taps into an aquifer.

Artesian water, mineral water, spring water, well water, sparkling mineral, and purified water are the categories of bottled water in the US. The definitions of each US bottled water category are as follows:

Wells that draw water from aquifers are used to harvest artesian water. These are subterranean aquifers made of porous rock or sand that, when tapped, release trapped water beyond the aquifer's level and occasionally even above its surface.

Mineral water comes from underground sources that are drawn at natural or bore exit points. Mineral water must include at least 250 parts per million (ppm) of total dissolved solids (TDS) and a constant concentration of mineral and trace elements at the point of emergence from the source. Keep in mind that ppm, which is the standard term for TDS on labels, is equivalent to milligrams per liter. It's also crucial to remember that adding minerals is not allowed.

Spring water refers to water that comes from an underground formation that naturally flows to the surface where it is collected or is alternatively tapped. The characteristic that sets spring water apart from mineral water—a steady mix of mineral and trace elements—does not exist. However, in actuality, their composition is almost always consistent.

Well water refers to any water that can be drawn out of a borehole or drilled into the ground that taps into an aquifer. "What is the difference between well and artesian water?" Although there isn't really a difference, some sources claim that there is because of the pressure aspect in artesian well systems, which causes water to rise to the top. Anyway, it is very uncommon to find bottled water labeled as well water.

Sparkling bottled water is defined as having the same amount of carbon dioxide in it as it did when it first emerged from the source, following treatment and the potential replenishment of any carbon dioxide that may have been lost during processing.

Any water that has been treated—even municipal tap water—by distillation, reverse osmosis, ozonation, deionization, or other appropriate methods and satisfies the criteria for "purified water" in the 23rd revision of the US Pharmacopeia is referred to as "purified water." One may also add minerals. Reverse osmosis or distilled water is recommended by the Centers for Disease Control and

Prevention (CDC) for people with compromised immune systems. According to a report by the nonprofit advocacy organization Food & Water Watch, as of 2014, more than half (64%) of all bottled water sold in the US was purified—that is, it came from the tap and underwent treatment (Bottled water vs. tap water, n.d.).

The primary distinctions between tap and bottled water are in source, choice, cost, and packaging (or lack thereof), not in quality or safety, which might change greatly between the two. A municipality's supply of tap water originates from a fixed surface or groundwater source, and its residents, for whatever reason, are not afforded a choice in the issue or the advantage of an alternative. They do, however, have inexpensive access to a plentiful supply of clean water. Although the cost of bottled water is significantly higher, customers have many options based on factors they may find significant, such as manufacturer, source, type, or taste.

The primary advantage of bottled water is its crucial function in giving low-income urban people in developing nations easily available, safe sources of water, and access to clean drinking water sources in an emergency. In this sense, the UN Sustainable Development Goal 6.1 (SDG 6.1), which aspires to provide everyone with equal and universal access to safe and inexpensive drinking water, will be greatly assisted by bottled water in the interim. For instance, 71% of Jakartans were dependent on bottled water in one way or another, according to a study that assessed the contribution of bottled drinking water to the achievement of SDG 6.1 in Indonesia.

The primary drawback of bottled water is its detrimental effects on the environment. Sustainability Harvard claims that using fossil fuels throughout the full life cycle of bottled water results in pollution and global warming. For instance, the Earth Policy Institute of Rutgers University reports that 86% of plastic water bottles used in the United States alone are not recycled and end up as trash or litter, even though the plastic (polyethylene terephthalate PET) that modern water bottles are made of is fully recyclable. Water that is bottled and packaged in plastic may also include obesogens, which are plastic-based compounds like BPA and phthalates that have been shown in studies to disrupt hormone balance and contribute to obesity (Bottled water vs. tap water, n.d.).

What Is Tap Water?

Water from a public water system that is provided by a water dispenser valve is known as tap water. Municipalities that oversee, manage, and operate treatment facilities or water sources that feed drinking water distribution networks are known as public water systems or PWSs. Over the past century, the creation and proliferation of tap water systems have completely changed the world by providing clean water to billions of people who would not have otherwise been able to escape the health and developmental issues stemming from decades of clean water shortage.

Public water systems are not all organized the same way worldwide. More than 90% of Americans receive their safe drinking water from the Environmental Protection Agency (EPA), which is the governing body in the US and works with the different states to fulfill this responsibility. There are three classifications of public water systems that are recognized by the EPA, namely Community Water Systems (CWS), Non-Transient Non-Community Water Systems

(NTNCWS), and Transient Non-Community Water Systems (TNCWS).

The same population of at least 25 individuals is served by the Community Water System (CWS) with water at their primary houses or at least 15 year-round primary residences. Water from the Non-Transient Non-Community Water System (NTNCWS) is provided to at least 25 individuals for six months out of the year, but not all year round. School buildings, hospitals, and corporate buildings are a few examples of NTNCWSs. Water from transient non-community water systems (TNCWS) is supplied to at least 25 individuals for 60 days or more a year; however, the individuals are not the same, and the supply is not consistent. Gas stations and campers, where people don't stay for extended periods of time, are examples of TNCWSs.

According to the EPA groundwater and drinking water statistics, approximately 155,693 public water systems existed in the US in 2007. Of them, 103,583 (66.5%) were non-community water systems, comprising 84,744 transient and 18,839 non-transient systems, and 52,110 (33.5%) were community water systems. Surface and groundwater are the sources of public water systems. According to the EPA, more people (70%) are year-round supplied by community water systems that use surface water from lakes, rivers, and reservoirs, even though the bulk of community water systems (78%) are supplied by groundwater (Bottled water vs. tap water, n.d.).

Apart from the previously discussed primary distinctions in choice and source, another noteworthy distinction is the distribution system. Only the tap water delivery systems from the source or water treatment plant to the service connection are regulated by municipal bodies. The homeowner/landlord is then in charge of maintaining and controlling the plumbing system that supplies the building with water.

With bottled water, there is no risk of external contamination or chain of responsibility because a single manufacturer maintains control over the entire process, from the point of origin to the bottling. That being said, there isn't much of a difference, particularly in industrialized nations, and according to Food & Water Watch, tap water is usually safer, cleaner, and healthier than bottled water. Naturally, this might not always be the case everywhere in the world, but generally speaking, it is. All around the world, tap water's primary benefit is its affordability. A gallon of tap water only costs pennies on the dollar. Tap water is 3000% less expensive per gallon than bottled water, costing between $0.02 and $0.64 per gallon, according to the Harvard Energy and Facilities department (Bottled water vs. tap water, n.d.).

The primary drawback of tap water is the need for a sophisticated infrastructure to provide proper regulation, upkeep, and contaminant control. The upkeep of a secure tap water system necessitates an efficient regulatory body that establishes and implements guidelines. Sadly, many developing nations lack the governmental framework necessary for this kind of organization, which partially explains why many of them lack access to clean tap water.

Moreover, tap water is more likely to be contaminated. Surface waterways, such as lakes, rivers, or reservoirs receive their feed from tributaries. Compared to bottled water, which is typically less exposed to environmental pollution, these surface waters may be subject to considerable pollution. The pipe networks that form the basis of tap water systems are frequently outdated or vulnerable to contaminant penetration. Tap water pollution can result from various factors, such as heavy metal leakage from ancient copper pipes, sewage, and environmental infiltration that enters the line from the outside.

Understanding Alkaline Water

The combination of hydrogen and oxygen is called water. That's why you refer to it as H_2O. The acidity of water is determined by its pH level, which goes from 0 to 14. Seven is regarded as the neutral pH number. It is said that the number "seven" is neutral or strikes a balance between acidic and alkaline. A pH of less than 7 indicates that water is "acidic." Higher than 7, it is considered "alkaline." According to EPA regulations, tap water's pH should be between 6.5 and 8.5. However, depending on where you live, tap water in the United States often has a pH between 4.3 and 5.3.

The regulations for bottled water vary based on whether it claims to be alkaline. The pH of bottled alkaline water is higher than 7. Sometimes, producers alter the water's chemical composition using a specialized tool. In other cases, they alter the pH of the water by adding nutrients. In contrast, black coffee has a pH of roughly 5, while orange juice has a pH of 3.3. Seven is the neutral pH of pure water.

Fans of alkaline water assert that, particularly after a strenuous workout, its higher hydrogen content offers better hydration than ordinary water. Adopters of the substance also assert that regular use of water with a pH lower than seven leads to excessive acidity in the blood and cells. Many health issues, including cancer and osteoporosis, are attributed to the low pH of ordinary water. More alkaline water is thought to lessen blood acid levels, enhance metabolism, boost vitality, reduce aging, enhance digestion, and lessen the loss of bone density, and proponents of high-pH water claim that it can starve cancer cells as well (Bonvissuto, n.d.).

These health claims regarding alkaline water are more attributed to sales than any scientific proof. Not much study has been done to back them up. Your body is capable of maintaining stable pH levels on its own. Your natural filtration system is your kidneys. Their role is to maintain the proper balance of acid in your body. Your body will expel more carbon dioxide from your breath if the acidity of your blood becomes too high. The great equalizer is your stomach. Your stomach secretions, which are made up of hydrochloric acid and digestive enzymes, neutralize everything you eat and drink. Basically, before alkaline water can accomplish whatever the health claims promise, your stomach will reacidify it.

Your tap water may taste different or bitter compared to alkaline water. Additionally, this can alter the flavor of anything you put it in, even tea or coffee. Alkaline water doesn't present any significant health dangers unless you have a kidney problem. The only other side effects of the high pH could be an upset stomach or dry, itchy skin.

Other Types of Water

Alkalized water does not seem to have much benefit, but when it is taken a step further and turned into ionized water, the benefits that it affords the human body skyrocket.

Ionized Water

In *The Miraculous Properties of Ionized Water*, Bob McCauley (2007) writes that the body has to be detoxified, alkalized, and hydrated for us to be healthy. We can prevent and even cure any ailment in the body if we succeed in doing this. We only need to figure out how to make this happen. All of these benefits and many more are obtained with ionized Water. There are three types of ionized water: alkaline (electrolyzed reduced water), acidic (electrolyzed oxidized water), and neutral ionized water.

Electrolyzed Reduced Water (ERW) has a pH that ranges from 7 to 12. Its redox potential ranges from -80 to -900 mV. It is a hydrogen-rich water, which endows it with antioxidant qualities. It functions mainly as a surfactant. Alkaline water actually has the ability to hydrate more quickly than tap water. More so than any other drug, it helps the body become more alkaline, hydrated, and detoxified. No other water has this ability. You can achieve a level of health you never thought possible by running regular tap water through a water ionizer. It's a miraculous process. Tap water is converted into electrolyzed reduced water, or hydrogen water. After filtering the tap water to remove contaminants like chlorine, the hydrogen and oxygen are separated using electrolysis. By donating electrons, this process effectively produces diatomic molecular hydrogen, also known as h_2 or molecular hydrogen. You are consuming antioxidant-rich water when you drink ERW because molecular hydrogen is a highly effective antioxidant.

By definition, an oxidant is an agent that oxidizes, whereas an antioxidant is a substance that acts to inhibit or stop the oxidation of another chemical. ERW can be used for drinking, cooking, cleaning, and cosmetics. Researchers in Japan have discovered that molecular hydrogen protects the cells' neurons and that ERW dramatically lowers oxidative stress-induced cell death. Many of the ailments that affect people today, including diabetes, heart disease, and cancer, have been linked to oxidative damage in the body. Conversely, antioxidants work to subdue the body's free radicals, or very reactive substances, which have the capacity to damage cells and result in illness.

Electrolyte Oxidized Water (EOW) is acidic water with a pH of 2 to 7. It can be quite acidic, but it can also be neutral. It has a redox potential of +400–+1200 mV. The high acidity in EOW does not offer as many benefits as ERW. Neutral ionized water is a substrate of EOW water. It is often referred to as weak acidity ionized water. EOW works better as a chemical disinfectant.

Deuterium Depleted Water

Water with a lower-than-normal concentration of deuterium is referred to as deuterium-depleted water (DDW) or light water. With a neutron adjacent to a proton, deuterium is a heavier isotope of hydrogen that is twice the mass of the hydrogen atom. Following the discovery of heavy water, it became evident that natural waters varied depending on the location, not only in terms of the number and composition of compounds soluble in water but also in terms of isotopic composition.

One of the two hydrogen isotopes that are twice as heavy as the standard isotope is deuterium. Unlike the normal isotope, which has zero neutrons, deuterium possesses one proton and one neutron. It is also referred to as heavy hydrogen and has an average density of one atom every 6,420 hydrogen atoms. Everywhere there is water, there is deuterium. In the world's seas, the concentration of deuterium is between 150 and 160 parts per million (ppm). In essence, deuterium exists in extremely minute amounts in all water.

Water that has either had its natural deuterium content reduced or removed is known as deuterium-depleted water. Deuterium naturally depletes in our bodies, maintaining a healthy level of about 100 parts per million. However, organisms may react negatively as they age or if they are regularly exposed to elevated quantities of deuterium through food, diet, and other external sources. Although deuterium is necessary for healthy cell growth, too much of it can harm cells or even hasten the creation of malignant cells.

Ocean average water (VSMOW) is a standard used to alter the isotopic composition of freshwater. The amount of deuterium in water is measured at 155.76 islands per million (ppm). Put another way, there is roughly one deuterium atom for every 6,420 hydrogen atoms in Earth's water. There isn't any real light water in nature with a deuterium level of 0 ppm. Antarctic snow contains the lightest water on Earth, which was created by atmospheric processes. Nature forbids us from consuming this drinking water because it is non-existent.

There are several ways to produce water that has been depleted of deuterium. It is feasible to deplete deuterium at home by partially freezing water and discarding the ice, albeit certain methods are more effective than others. It takes a lot of time, and the outcome varies based on the surroundings. When water vapor evaporates from the earth's surface, lighter hydrogen rises, leaving the heavier deuterium concentration behind. This is how deuterium depletion occurs naturally through the water cycle. This helps to explain why the majority of deuterium is found in oceans. As a result of this process, deuterium is discovered less frequently at higher elevations and farther from the coast. A number of techniques are available for scientific deuterium depletion methods. Fractional distillation is a popular but slow and expensive method. Other techniques include the isotopic vacuum distillation technique.

Water Assessment

The human body is made up of extracellular substances and live cells arranged into organs, tissues, and systems. In this subchapter, we will compare bottled water, tap water, alkaline water, ERW, and DDW by looking at how they affect certain aspects of the human body.

Cells

The body's fundamental building blocks are called cells. Each and every tissue and organ is composed of billions of distinct cells. Although they differ in size, all human cells are quite tiny. Not even the biggest, a fertilized egg, is visible to the naked eye. Every cell in an individual's body is descended from two cells: the sperm and the egg. The fertilized egg is merely one solitary cell once the sperm and egg combine (fertilization). The zygote cell divides repeatedly, giving rise to offspring cells with distinct traits and roles with each division. The many organs are ultimately formed by these various cells. Numerous distinct cell types, each with a unique structure and function, make up the body. Among the several cell types are blood cells, muscle cells, skin cells, nerve cells, and glandular cells. Certain cells are not connected to one another and can flow freely in the circulation, such as blood cells. Muscle cells and other types of cells have strong connections to one another. Skin cells are among the cells that proliferate and divide swiftly. Certain types of cells, including nerve cells, don't divide or proliferate unless specific conditions are met (Villa-Forte, 2022).

We've already touched on how DDW and ERW affect the body at the cellular level. No other water types are able to do this. While tap water and bottled water are able to help transport nutrients to different parts of the body, DDW and ERW are able to do this more efficiently. These waters hydrate the body much better than all the other waters.

Tissue and Organs

Tissues are collections of connected cells. Although the cells that make up a tissue are not all the same, they cooperate to carry out certain tasks. For instance, the muscle cells in muscle tissue contract to produce muscle movement. Nerve cells in muscular tissue also transmit signals that instruct muscles on when to contract and relax. The strong, frequently fibrous tissue that gives the body's structures strength and suppleness is known as connective tissue. It makes up a significant portion of the skin, tendons, joints, ligaments, blood vessels, and muscles, and it is found in practically every organ. Depending on where in the body it is located, connective tissue has different properties and cell kinds.

The identifiable bodily parts that serve particular purposes, such as the stomach, liver, eyes, heart, and lungs, are called organs. An organ is composed of several tissue types and, consequently, various cell types. The heart, for instance, is made up of fibrous tissue that forms the heart valves, muscular tissue that contracts to pump blood, and unique cells that regulate the pace and rhythm of heartbeats. The eye is made up of nerve cells that send impulses to the brain, muscle cells that control pupil opening and closure, clear cells that form the cornea and lens, cells that generate the fluid inside the eye, and cells that detect light. Even an organ as seemingly simple as the gallbladder is made up of various cell types, including muscle cells that contract to evacuate bile, cells that make the fibrous outer wall holding the sac together, and cells that form a lining resistant to the irritating effects of bile (Villa-Forte, 2022).

In Chapter 6, we discuss fascia in greater detail. ERW and DDW affect the fascia unlike any other water type. Both help the body's detox systems to function optimally. The reason most of us feel the need to detox our bodies is that bottled, tap, and alkaline water are not able to handle the demands on our bodies. DDW and ERW are easily absorbed into the system, which means that you don't have a lot of waste from inefficiencies.

Organ Systems

Organ systems are collections of organs that work together to perform collective functions, even if each organ has a distinct purpose. Physicians group illnesses and their own clinical specializations based on organ systems. The digestive, cardiovascular, and musculoskeletal systems are a few instances of organ systems and their roles. The digestive system, also known as the gastrointestinal tract, is in charge of food intake, breaking it down, and eliminating waste. It runs from the mouth to the anus. Along with the stomach, small intestine, and large intestine, which move and absorb food, this system also contains related organs, including the liver, gallbladder, and pancreas, which create digestive enzymes, eliminate toxins, and store materials needed for digestion.

Organ systems frequently cooperate to complete challenging tasks. For instance, following a substantial meal, multiple organ systems cooperate to support the digestive system's need for extra blood to carry out its operations. The neurological system and the cardiovascular system assist the digestive system. The digestive system's blood vessels enlarge to carry more blood. The brain receives nerve impulses, alerting it to the increased activity of the digestive system. Through the release of substances into the circulation and nerve impulses, the digestive system even directly stimulates the heart. In response, the heart pumps blood more vigorously. As a result, the brain perceives less hunger, more fullness, and less desire for strenuous physical activity (including the musculoskeletal system), saving more blood for the digestive system instead of the skeletal system.

Organ systems and organs must communicate with one another. The body uses communication to modify each organ's function to suit the needs of the entire body. In the aforementioned example, the heart needs to know when to pump extra blood to the digestive organs. The heart can pump less when it recognizes that the body is at rest. In order for the kidneys to preserve water and generate more urine when the body is dehydrated, they need to be aware of when the body is overly hydrated. The process by which the body preserves its typical composition and operations is referred to as homeostasis. The body can keep internal fluid and material levels steady because organ systems communicate with one another. Furthermore, none of the organs overwork nor underwork, and each organ supports the operations of the others. Endocrine and autonomic nerve systems communicate to keep the body in homeostasis. The communications are carried out via unique substances known as transmitters.

The intricate network of communications that governs how the body functions is mostly controlled by the autonomic nervous system. This portion of the nervous system runs without the user's awareness and without producing any overt symptoms. Neurotransmitters are a type of transmitter that facilitates communication inside the nervous system as well as between the nervous system and other organs. The hormones that are produced by the many glands that make up the endocrine system are chemical transmitters. Through the bloodstream, hormones reach other organs and control how well they operate. The thyroid gland, for instance, is responsible for producing thyroid hormone, which regulates the metabolic rate—the rate at which the body's chemical processes occur. Insulin is produced by the pancreas and regulates how sugar is used (Villa-Forte, 2022).

A class of steroid hormones known as estrogen is known to support the upkeep and development of features unique to the female body. Estradiol is one of the forms of estrogen that can be found in tap water. This is the strongest form of estrogen that can get into the water system from contraceptive pills that women pass through their feces or urine. Chronic hormone use has been linked to health risks, including potential effects on prepubescent teens' sexual development. For these children's bodies to develop the proper sexual traits within the designated time, the availability of controlled levels of male and female hormones is crucial. Alkaline water does not balance sex hormones, although it does balance out bone density in older people. DDW and ERW do not have high levels of Estradiol.

Barriers on the Inside and Outside

As bizarre as it may sound, the body's multiple surfaces make it difficult to always tell what is within and outside it. Of course, the skin is an organ system and is located outside the body. It creates a barrier that keeps a lot of dangerous things out of the body. Beginning at the mouth and winding its way through the body, the digestive system is a lengthy tube that ends at the anus. Is food entering or leaving the body as it travels through this tube? Until they are absorbed into the bloodstream, nutrients, and liquids are not truly internal to the body.

Air enters the trachea, the windpipe that leads from the nose and neck to the bronchi, the large network of branching airways that make up the lungs. When does this internal channel occur in the body? The body cannot use oxygen in the lungs until it gets into the bloodstream. Oxygen needs to get past the lungs' thin cell lining in order to enter the bloodstream. This layer serves as a barrier against bacteria and viruses that can enter the lungs through the air, such as those that cause tuberculosis. Most airborne infectious organisms never cause disease because the lungs have numerous defense mechanisms, such as antibodies to combat infection and microscopic hairs called cilia to clear debris out of the airways.

In addition to dividing the interior from the outside, body surfaces also maintain materials and structures in their correct locations to allow for optimal performance. For instance, because blood is generally contained in blood vessels, internal organs do not float in a pool of blood. Not only does blood fail to supply oxygen and nutrients to tissues when it leaks from the veins into other areas of the body, but it can also cause serious damage. For instance, because the skull cannot enlarge, even a tiny amount of blood seeping into the brain can kill brain tissue. However, because the abdomen has the capacity to expand, a comparable volume of blood flowing into it does not induce tissue destruction (Villa-Forte, 2022).

When our guts are not working how they should, alkaline water can be helpful in maintaining our pH levels. Acidity outside of our guts can be disastrous. The systems discussed in this subchapter all contribute to detoxification. DDW and ERW manage the payload of toxins so that our systems are not overloaded.

Overall, we can see that DDW and ERW come out on top when it comes to the advantages they bring to our bodies.

Chapter 3:

Benefits of Electrolyzed Reduced Water and Deuterium Depleted Water

In Russia, negative ions were initially explored in the 1850s. They are among the most fundamental and common components of human health. The body cannot survive without negative ions. Biological aging is the result of positive ions, which gradually deteriorate our health until we pass away from illness and our organs stop working. Most elderly people pass away quietly, and the common phrase used to characterize this type of death is "died of old age." In reality, it implies that so many of their organs have failed to work that it is hard to identify the cause of their death.

Getting enough negative ions into the body is a crucial part of the battle for good health. Eating raw food and consuming ionized water help achieve this. Living in a positive ion environment both inside and outside the body is detrimental to our health. Given this, it is impossible to overstate how important the discovery and research of negative ions is to the advancement of human health.

Beginning in the early 20th century, a large portion of the preliminary research and creation of ionized water was conducted by Russians. Through their discovery of negative ions and the research they conducted on them, they concluded that since negative ions were beneficial to human health, positive ions must also be detrimental, which is really the case. Additionally, they hypothesized that water might acquire some very healthful qualities if they could introduce negative ions into it. They experimented with basic electrolysis, which involves simultaneously submerging an anode and a cathode in water to produce positive and negative ions that cancel each other out.

How to separate the ions was the problem they had to solve. Early in the 20th century, the Russians invented the first water ionizers. Though it has been largely unacknowledged until recently, ionized water is one of Russia's greatest scientific triumphs and a wonderful gift to the world. In the 1940s, more ionized water tests were conducted in Russia (then the Soviet Union). The goal of the Chelyabinsk Project, which was founded in the late 1960s, was to ascertain whether ionized water could successfully eliminate radiation from the body. Under the direction of Vladimir Egov, the Chelyabinsk Project was named after the little village where it happened in the Ural Mountains. Radiation exposure occurred to a large number of people in the former Soviet Union during Cold War nuclear research and development. Thus, the need for techniques to eliminate radiation from the body was urgent. Russia introduced three methods (Chlorella Pyrenoidosa, Focoidan, and Far Infrared Waves) to remove heavy metals from the body; processes that were made easier by regularly consuming ionized water because hydrated tissue detoxifies more readily than dry tissue.

In the 1950s, Japan produced the first water ionizers. During this period, some of the earliest experiments were conducted on plants and animals. The Japanese medical community concluded very quickly that Ionized Water did not pose a health risk to humans. In 1958, the first water ionizer for domestic use was created and distributed. The Japanese Health Ministry certified the water ionizer as a "Health Improving Medical Device" on January 15, 1966. Ten years later, Korea followed suit and authorized the water ionizer as a medical device (McCauley, 2006).

Ionized water research and development continued in other nations during the 1970s and 80s, mainly in Japan. It marked the start of the water ionizer's domestication. It became simpler to get healthier, and our understanding of what constitutes healthy water would also change.

Ionized water is frequently called "living water," and DDW water joins those ranks as well. In this chapter, we expound on what sets these two glasses of water apart by considering the science behind how they work, how they supercharge detoxification and all the other ways they provide benefits to the body.

The Science Behind Electrolyzed Reduced Water

Water electrolysis is an inventive technology used in the creation of ERW, as expressed earlier. The cathode and anode are isolated in this particular process using a particular membrane. Whereas the anode generates acidic water (EOW), the cathode will yield electrolyzed reduced water. Since the membrane is semi-permeable, neither of these waters can mix. This explains why the pH values in each of the fluids are maintained. Many therapeutic benefits are exerted by the hydrogen gas content of ERW.

ERW is another term used by chemists to refer to alkaline water with antioxidant potential. Any tap water with a pH higher than 7 has no potential antioxidants. It cannot, therefore, be regarded as an ERW. It's time to acknowledge that not all alkaline water has the same capacity to produce antioxidants. Antioxidant qualities are present in the alkaline water generated by an alkaline water ionizer. Its value in terms of health benefits may be unique. ERW is produced as a consequence of a redox reaction.

The simultaneous occurrence of oxidation and reduction is what is referred to as redox. A large number of electrons are gained by the material during the reduction phase. It has the power to oxidize the material that has experienced reduction. The substance begins to lose electrons during the oxidation phase. It improves one's capacity to take in electrons from other substances.

The electrons travel throughout the redox reaction, both acquiring and losing energy. The electrolyzed reduced water's oxidation-reduction potential (ORP) is the outcome of this. Voltage is used to express potential. Negative ORP is included in the substance that loses electrons during the redox reaction. On the other hand, the one that takes in electrons has a positive ORP. Since the entire purpose of ORP is to measure electrical energy potential, it's expressed in millivolts.

Reactive oxygen species (ROS) and oxidation-reduction potential (ORP) differ slightly. The ORP quantifies how well an alkaline water ionizer's water can lessen the harm caused by that reactive oxygen species (ROS). ROS can cause the greatest amount of damage to tissues and DNA through oxidation. It might be the result of free radicals produced by an antioxidant-rich diet. They cause the tissues to corrode negatively. The damage brought on by ROS may be completely stopped by the antioxidants. The body's response helps prevent additional harm to the tissues and DNA. Consuming decreased water that has been electrolyzed makes it feasible.

Ionized alkaline water is the colloquial term for electrolyzed reduced water. It's a water ionizer that assists in giving the water a negative electrical charge, which produces a reduction potential. Water with low reduction potential can aid in oxidative neutralization. The body's ability to neutralize reactive oxygen species (ROS) is closely correlated with the reduction potential of ionized alkaline water. Alkaline water's reduction potential results from its antioxidant capacity. It has been established that one of the most prevalent reactive oxygen species in the human body is hydrogen peroxide. Reactive oxygen species within the body can be neutralized by the electrolyzed reduced water.

Hydrogen peroxide is reduced to water and hydrogen with the aid of the ionized alkaline water generated by a functioning water generator. It may also facilitate a hangover's quick recovery. Excess hydrogen peroxide accumulation causes serious harm to the liver tissue. A potent antioxidant can be made by combining vitamin C with ionized alkaline water. Vitamin C's antioxidant capacity is increased when it is combined with alkaline water. Consuming the mixture rapidly will contribute to the instantaneous destruction of antioxidant beverage advantages. Electronically reduced water is very well-liked due to its potent antioxidant properties. It has outstanding health benefits as well (Fujiiryoki Purifier, 2023).

Detoxification Benefits of Deuterium-Depleted Water

Your body is equipped with all it requires to cleanse itself. As you read this, it has already begun the detoxification process. How effectively is it accomplishing that, though? That depends on your general state of health. Enhancing your body's detoxification systems is essential to bettering your health. When you take care of your detox organs, your general health will improve. Advertisements attempting to promote "detox" as a neatly packaged product or a seven-day crash diet regimen are everywhere. In actuality, your body's internal organs are responsible for constantly cleansing your body. It's no accident that the detox organs are frequently called "the vitals." Your detox organs, like your immune system, are unseen and frequently underappreciated but always on the defensive, protecting your body.

Your body's continuous cleansing activities are a manifestation of its vitality, which shows itself as excellent health, much like your heart's continuous pounding. Your health will rapidly deteriorate if your detox organs are not operating at optimal capacity at all times. Thus, your "detox plan" should consist of getting to know your detox organs well, comprehending how the detox processes operate, and determining which organs need special care to maximize these processes. You will benefit greatly from this detox plan for the remainder of your life. This is just a healthy way of living, based on self-awareness and a personalized approach, not a marketing plan.

There will be times when different strategies are required to put this strategy into action. In general, they will involve giving specific organs some mild, natural support to ensure that the delicate detoxification process proceeds smoothly and as nature intended. Your skin, kidneys, lungs, and liver are all made to continuously detoxify you. Although these organs serve a variety of purposes, their main duty is to maintain your life and well-being by working constantly throughout the day.

Every choice you make regarding your detox should be based on these essential organs and how they are supposed to aid in your body's detoxification. One of these organs may be impacted by a health issue that while appearing unrelated to detoxification processes, could impair the organ's capacity to cleanse your body. Additionally, health conditions like skin disorders that may appear to be superficial and unrelated to the body's major organs could actually be an indication that your body's detoxification systems require review.

The Lungs

Your body is being detoxified while you breathe. Although the liver is often referred to as your body's detox superhero, the number of breaths you take and exhale in a day—between 15 and 25 every minute—serves as a clear reminder of the critical role your lungs play in maintaining your vitality. The lungs, which comprise the bronchi and respiratory tract, do, in fact, have their own filtration system and are designed to play a vital role in your body's continuous detoxification activities.

Your lungs are expected to continuously filter out carbon dioxide waste (in the form of poisonous gas) and ensure that every single cell in your body has access to adequate oxygen to survive, while your liver and kidneys are meant to be busy filtering your blood and preventing poisons from entering your system. This is the fundamental mechanism that underlies breathing, which functions as a detoxification system to remove poisons from the body while retaining healthy molecules. However, it is so essential to your survival as a detox process that it not only occurs continuously but is engineered to do so in a nanosecond or less.

Your Lungs are designed to assist in the detoxification process if the Liver, Kidneys, and intestinal tract (a portion of the stomach and digestive system) are unable to get rid of a toxin or break it down so that it may be completely digested and expelled. The bloodstream absorbs the problematic poisons or compounds and carries them upward, where they enter the alveolar sacs that make up the base of your lungs. The tough toxins are supposed to be taken up by the lungs and subsequently coughed up as phlegm.

Cigarette smoke can overwhelm the cilium, the lung's filtering mechanism, which can eventually result in a reduction in the lung's ability to operate as a detoxifier. This is one of the reasons smoking tobacco is discouraged.

The Skin

The skin serves as your body's first line of defense against pathogens and triggers your immune system when dangerous bacteria manage to get through. However, in addition to keeping healthy chemicals in your body, the skin is also essential for eliminating toxins from the inside. The primary role of the skin in the detoxification process is sweating off waste products from digested food that are high in protein, refined sugar, or acidic, as well as urea and uric acid.

Your skin naturally uses a unique enzyme called cytochrome P450 to transform a wide range of undesirable substances into a form that is water-soluble. If this is done correctly, the chemicals are now transformed into groups of toxins and are expelled from the body through the pores in the skin and the sebaceous glands, which produce sweat. The skin's function is to collect toxins and release them through sweat when the body's detoxification systems—the liver, kidney, and lungs—become overwhelmed by the volume of waste that needs to be removed.

But keep in mind that there is a price for this detoxification phase: skin issues. In the event that the liver, kidneys, and lungs are unable to handle the toxins, the skin will attempt to eliminate them from your body. Skin obstructions brought on by such a large amount of toxic burden appear as boils, rashes, cysts, blackheads, and whiteheads on the skin's surface. Treating internal organs and optimizing their detoxifying capacities is a better approach for treating skin disorders like acne than focusing all of the care on the skin's surface.

However, sweating toxins out of your body is still a normal part of the continuous natural detox process. In fact, we highly recommend that you use daily physical exercise to facilitate this process. Achieving a state of exhaustion and perspiration can stimulate your circulatory system and enhance the removal of toxins from your body by pushing them to the skin's surface (*The Inside Workings of Your Body's Detox Explained*, n.d.).

The Liver

Your body's complete system of detoxification is primarily handled by your liver. The liver's job is to protect you against poisons found in food additives, over-the-counter medications, alcohol, and other addictive substances while preserving the healthy elements your body needs to function. The liver should receive the substances you consume and dissect them into their constituent chemicals as part of this detoxification process. The purpose of your liver is to store the good chemicals and transfer them throughout your body through the bloodstream, which first flows through your liver.

The dangerous, poisonous substances are intended to be removed from your bloodstream and further broken down by your liver, preparing them for the subsequent phase of processing that will ultimately result in elimination. It is important to keep in mind that constipation can impair the Liver's capacity to cleanse your body. This is due to the fact that constipation causes waste products that need to be expelled from the body to remain in the colon. These waste products eventually get reabsorbed, which forces the liver to undergo repeated detoxification, depleting its vital energy (*The Inside Workings of Your Body's Detox Explained*, n.d.).

The Kidneys

By filtering your blood and removing waste products (drugs, ammonia, urea, and poisonous compounds) from it, your kidneys continue the liver's detoxification process and finally produce urine that is ready for expulsion. One of the most crucial phases of your body's detoxification process is keeping your bloodstream clear of waste and extra fluid, a job handled by your kidneys. Water consumption is very important for your kidneys to perform this function correctly. Your kidneys should be able to filter out waste and dangerous substances from your bloodstream, cleansing it in the process if you are consuming the recommended amount of water for your particular situation.

Your kidneys should filter your blood 20 to 25 times a day on average. Pressurizing the blood as it passes through the intricate internal structure of the kidneys, which resembles a sieve, is a continuous process. The detoxification activities of your kidneys might be severely compromised by dehydration. Your blood requires a specific amount of water to thin it out to travel through the kidney's structure during the filtering process.

However, your blood pressure readings are just as crucial. High blood pressure, or hypertension, can cause irreparable damage to the kidneys and renal failure. These days, hypertension is one of the leading causes of renal failure worldwide and a major global health concern. In England, thousands of people suffer from excessive blood pressure, albeit many are not aware of it. Your kidneys' vital role in controlling blood pressure makes them particularly vulnerable to hypertension, even though they are made to handle a large amount of the daily detoxification process. Chemicals that are hard to digest can also clog and damage the kidney's natural membrane structure, which resembles a sieve (*The Inside Workings of Your Body's Detox Explained*, n.d.).

The cell is the basic building block of our bodies. When cells are clustered together, they become tissue. When tissues cluster together, it becomes an organ. Our organs are made up of cells, and every cell has mitochondria. DDW water affects the health of our mitochondria so that our cells function the way they are supposed to. It should, therefore, come as no surprise that DDW affects cancerous cells. Cancer occurs when cells stop doing what they are supposed to do, and every organ involved in our detoxification system can go rogue and become cancerous. DDW can, therefore, improve the body's natural detoxification processes, making sure that all the cells function optimally.

How These Waters Can Improve Overall Health

Much like water, heavy water is used by the human body, but it also kills the body's chemical interactions. The most basic energy response in the majority of organisms, the mitochondrial respiratory chain response, is inhibited by heavy water molecules. Consequently, there is a reduction in the amount of side and cell energy utilized, which can have detrimental effects on the body and raise the chance of illness. It is now established that the presence of heavy water in the human body can lead to a number of illnesses, such as type 2 diabetes, depression, and infectious disorders brought on by compromised immune systems.

When two liters of deuterium water (125 ppm deuterium) are consumed daily, the amount of deuterium in the human body can change by less than 1%. Adults have between 120 and 140 parts per million of deuterium in their bodies. Even if it seems like such a little sum, utilizing light water raises the energy level in living things and enhances the energy in already-existing cells. These extra resources allow organisms to address possible issues in their essential functions.

The following are the effects of consuming light water: delaying the onset of aging, boosting the immune system, keeping skin feeling fresh, detoxification, adjusting the digestive system, building muscles, increasing vascular response, and improving tumor cells.

The slower kinetic energy production in cells is one of the primary links between this heavy isotope and the health of your cells. In essence, lighter hydrogen is needed by our cells to perform chemical processes quickly and effectively. This process is slowed down by a higher deuterium content, which affects important processes like DNA replication and repair. The tiny nano-motors called mitochondria are found inside our cells, where they provide the energy and metabolic water required for life. These nano-motors run on hydrogen, but as deuterium levels rise, the increased viscosity acts as a brake, reducing the amount of energy and metabolic water produced by the motors. We grow more and more exhausted as the influence on nano-motors increases. Deuterium-depleted water has been shown to improve cell function, slow the growth of malignant cells, and improve sleep cycles and energy levels.

Elevated deuterium levels affect mitochondrial activity, the organelle responsible for burning fat and carbs in your body. Therefore, maintaining healthy mitochondria is essential to preventing a number of illnesses, including diabetes, weight gain, mental health issues, and abnormal brain function. Reducing the amount of deuterium in the water we consume can have a significant effect on our cells' health and general well-being (Govender, n.d.).

ERW benefits the body in a similar way. The ability of ionized alkaline water to prevent cancer is undoubtedly one of the greatest perks. Oxygen radicals are a byproduct of several vital metabolic processes and are responsible for a great number of malignancies in the body. These unstable molecules, known as radicals, cause disturbances in the body's cells that have the potential to develop into cancer. Antioxidants, which an ionizer produces in the form of Hydroxide ions (OH-), can counteract these hazardous free oxygen radicals. These ions function similarly to other beneficial antioxidants like vitamins A and C. They can join forces with oxygen radicals to form a complete oxygen molecule, which the body can use realistically.

Not only can free radicals cause cancer, but numerous studies have also linked them to worsening cardiovascular health over time. They have specifically been connected to atherosclerosis, which is the build-up of cholesterol in the inner vessel wall caused by damage. This is frequently a precursor to a heart attack or stroke because a sufficiently large atheroma can narrow the blood vessels and restrict blood flow. Additionally, fragments of the atheroma may break off and become lodged in a vessel, resulting in a heart attack.

By initially neutralizing free radicals and stopping them from attacking vessel cells to initiate the formation of atheroma, alkaline ionized water can reduce these dangers. Curiously, it has also demonstrated some encouraging outcomes in diminishing the possibility of an atheroma developing and lessening its danger even in the event that one does. Additionally, ERW is more hydrating. This can seem strange at first because, obviously, drinking regular tap water is a terrific way to stay hydrated. Nevertheless, the ionization process disperses the water's molecular clusters into smaller groups. Compared to normal clusters, which normally include 10 to 15, this procedure leaves a greater proportion of clusters that contain just about 5 H_20 molecules.

Because there is less resistance to these little clusters, they can pass through membranes more quickly and be absorbed into cells more easily. As a result, we become more hydrated quicker and may practically use water molecules to clean out waste from our bodies.

It has also been demonstrated that alkaline ionized water has an interesting effect on digestion and is especially advantageous for the colon. Because the colon is a vital component of the digestive system, it must be maintained as alkaline as possible. This not only makes it work better, but it also lessens the chance of developing colon cancer.

It's also critical to understand that consuming alkaline water doesn't significantly worsen the stomach's digesting process. Your body would still be able to digest food readily even if the acid were to become slightly less strong because enzymes, not the acid itself, perform the majority of the digestion process. However, it is usually advisable to wait 20 minutes after eating to consume any alkaline water if you are eating meats, lactose, or gluten. As an example, Kangen Water® has been used in Japan to help restore the body to its alkaline state.

It is obvious that drinking ERW provides numerous significant health advantages. ERW, however, also contributes to bright, fresh, wrinkle-free skin, so its advantages are not limited to treating serious medical diseases. It achieves this by using the hydrogen ions generated by ionization to attack free radicals. Since these radicals harm the skin and cause it to degrade over time, they are the primary source of wrinkles. Therefore, it is crucial to combat and neutralize them, which is commonly accomplished with countless facial exfoliators and moisturizers. Rubbing ionized water on the face has also been linked to lower levels of acne and fewer painful spots (7 Benefits of Water Ionizers, n.d.).

Chapter 4:

Sourcing the Best Water for Our Homes

The Journal *Pesticides and You* shared the story of a groundskeeper, Steve Herzog, who exposed drinking water contamination. I am going to let Herzog (2011) tell his story:

This is a first-hand narrative from me, Steve Herzog, a cancer survivor, whistleblower, and groundskeeper who prevented the Yale University Golf Course (YGC) from using its contaminated drinking water source. I was a groundskeeper at YGC from 1983 to 1996. Many pesticides, herbicides, and fertilizers were used when I was there, disregarding the fact that YGC uses well water for drinking and is close to Maltby Lakes, a recreational area.

When the Yale University Office of Environmental Health and Safety tested the cold water from a faucet in the maintenance barn in August 1989 and discovered 3 coliform bacteria per 100 milliliters, I got concerned. I then attempted to have YGC test the well water for particular fertilizers and pesticides, but I was informed that it would be too costly. Naturally, this raised my suspicions because it was clear that YGC had the resources to conduct any testing required to guarantee the well water's safe condition.

Another reason I was worried was that one of my coworkers had "the shakes"—his head and upper body would shake constantly, and his skin was yellow. He eventually passed away, and I discovered much later that yellowing of the skin and "the shakes" are symptoms of high coliform exposure.

During my 13 years on the YGC Safety Committee, I attempted, in vain, to persuade those in authority to limit the use of certain fertilizers and pesticides. The one exception was when, through years of discussions and research that I spearheaded, I convinced YGC's upper management to outlaw the use of Milorganite, a fertilizer made from sewage sludge that is known to contain heavy metals like arsenic and mercury.

But the superintendent had me use a walk-behind fertilizer spreader to apply it to all the greens and tees. It appears that this was my punishment for my efforts at the time. After that, I experienced nosebleeds for weeks on end because I had no protection and was informed that it was "organic and safe." You might wonder why I didn't say no to the task. I would think to myself in the back of my mind, "This is Yale University, one of the best universities in the world." It needs to be secure to use.

When the Safety Committee met in the YGC clubhouse in October 1995, I recall becoming extremely irate and outspoken during a discussion about using Calo-Chlor, a fungicide based on mercury that has since been phased out, to manage snow mold on the greens over the winter. The new superintendent that YGC appointed stated, "I am the superintendent, and I will use whatever I want to" at this meeting. I loudly voiced my concerns.

I felt quite weary and found a bump on my neck a few weeks after this meeting. It was discovered to be cancer. In January 1996, Yale Pathology initially diagnosed me with anaplastic carcinoma. However, in October 1997, my occupational physician ordered fresh stains to be run on the initial biopsy, which led to the diagnosis of non-Hodgkins lymphoma. With a good degree of medical certainty, the occupational doctor decided that I developed lymphoma as a result of being exposed to several pesticides at YGC, including well water.

My occupational doctor came to the conclusion that chemicals that leaked into the well system after being used to wash mowers and pressure-treated wooden steps that contained arsenic were the sources of the contamination in the well water. After receiving my complaint, the Environmental Protection Agency (EPA) in the United States opened a criminal investigation. The water quality division of the Connecticut Department of Public Health discovered that the well water was tainted with p-dichlorobenzene (3.0 ug/l) and nitrates (12.4 mg/l), both of which are harmful to human health.

It should be mentioned that I filed an anonymous letter to Connecticut's then-attorney general, Richard Blumenthal, to expedite the process because the EPA's Criminal Probe Division probe was taking so long. As of the end of January 2001, YGC was linked to the municipal water supply of the City of New Haven due to the EPA inquiry. I recently found out that Maltby Lakes hasn't been used for drinking water since the early 1980s. But throughout my time at YGC, I was consistently informed that Maltby Lakes served as New Haven County's backup reservoir. I pray and hope that my story prevents contamination and poisoning at other golf courses! I have been cancer-free for fifteen years, and the

Yale University golf course has stopped using its tainted drinking water!

Unfortunately, this is not an isolated incident. Water contamination happens quite often. This means that drinking water straight from the tap may be detrimental to us. Pesticides are only just one problem. There are a variety of contaminants that we should be aware of, which will be one of the focal points of this chapter. We will also delve into the regulation of our drinking water and the steps that we all can take to improve its quality.

Water Quality Standards and Regulations

In drinking water, the EPA establishes regulatory limits on more than 90 contaminants. A contaminant's legal limit is the highest amount that can be achieved by water systems with the finest technology while still protecting human health. Water systems must abide by the EPA's water-testing schedules and procedures. States may establish and implement their own drinking water standards under the Safe Drinking Water Act (SDWA), provided that the criteria are at least as strict as the federal standards imposed by the EPA.

National Primary Drinking Water Regulations

The National Primary Drinking Water Regulations (NPDWRs), created by the EPA, impose obligatory water quality criteria for pollutants found in drinking water. These legally binding requirements, known as "maximum contaminant levels" (MCLs), were put in place to safeguard the general public from ingesting toxins in drinking water that could endanger their health. The maximum permitted level of a contaminant in drinking water that is supplied to the customer is known as an MCL. Public health is safeguarded by primary standards and treatment methods that reduce the number of pollutants in drinking water: microorganisms, disinfectants, disinfection byproducts, inorganic chemicals, organic chemicals, and radionuclides.

National Secondary Drinking Water Regulations

Furthermore, the National Secondary Drinking Water Regulations (NSDWRs) specify non-mandatory requirements for water quality for fifteen different contaminants, all created by the EPA. These "secondary maximum contaminant levels" (SMCLs) are not enforced by the EPA. They serve as recommendations for public water systems to manage the aesthetics of their drinking water, including color, odor, and taste. At the SMCL, these pollutants are not thought to risk human health.

The EPA mandates a special notice for any exceedance of the fluoride SMCL of 2.0 mg/L, even though SMCLs are not legally enforceable. Community water systems that are above the SMCL of 2 mg/L for fluoride but not the MCL of 4.0 mg/L are required to notify the public of the exceedance within 12 months after the water system's discovery of the exceedance (40 CFR 141.208).

The SMCL public water systems merely need to conduct voluntary tests for these chemicals, which pose no health risks. What makes setting secondary standards important, then? According to EPA estimates, these pollutants may give your water a poor taste or smell and make it appear murky or discolored if they exceed certain levels. Secondary standards are designed to provide public water systems with some direction on reducing these contaminants to amounts that are unlikely to be evident to most consumers. Secondary pollutants are linked to a broad range of issues.

Three categories can be used to classify these issues (CDC, 2023):

- Aesthetic effects: undesirable tastes or odors (Chloride, Copper, Foaming Agents, Iron, Manganese pH, Sulfate, Threshold Odor Number (TON), Total Dissolved Solids, Zinc)
- Cosmetic effects: undesired outcomes that do not harm the body (Silver, Fluoride)
- Technical effects: deterioration of water equipment or decreased efficacy of other pollutant treatment

(chloride, copper, corrosivity, iron, manganese, pH, total dissolved solids, Zinc, aluminum)

Unregulated Contaminants

The drinking water CCL is a list of contaminants that are known or expected to exist in public water systems but are not yet covered by any planned or issued national primary drinking water standards. Under the Safe Drinking Water Act, contaminants included in the CCL may need to be regulated in the future (SDWA). Every five years, the EPA is required by SDWA to publish the CCL. When deciding which unregulated pollutants to list, the Agency is directed by the SDWA to take into account the health consequences and incidence data. Additionally, the SDWA mandates that the Agency designate the contaminants that pose the biggest risk to public health when ingested through drinking water. In order to determine priority pollutants for regulatory decision-making, the EPA uses the CCL.

In order to determine which uncontrolled contaminants should be taken into consideration for the CCL and which may necessitate a national drinking water regulation, EPA has solicited the public for information on contaminants starting with CCL 3. Early public participation in the CCL development process is something both the National Academy of Sciences (NAS) and the National Drinking Water Advisory Council (NDWAC) encourage.

In a different procedure known as Regulatory Determinations, the EPA must decide whether or not to regulate at least five pollutants from the final CCL once it is released. If more information is available, EPA will gather it and assess it for each of the CCL pollutants to identify which ones have enough details to be assessed in accordance with the three SDWA criteria before deciding which regulations to apply. For the five or more CCL contaminants for which there is enough information to compare them to the three SDWA criteria, the EPA will determine the appropriate regulatory action. For contaminants that don't have enough data to support a regulatory decision, EPA will keep gathering data, carry out and support research, and find ways to close data and information gaps.

A regulatory determination is a formal decision on whether EPA should initiate a process to develop a national primary drinking water regulation for a specific contaminant. The law requires the EPA to make regulatory determinations for at least five contaminants from the most recent CCL within five years after the completion of the previous round of regulatory determinations.

It is important to remember that the EPA is not confined to determining regulations for just the pollutants listed on the CCL. If data becomes available indicating that a particular contamination poses a risk to public health when present in drinking water, the Agency may also choose to regulate other unregulated contaminants (CDC, 2023).

Preventing Waterborne Germs at Home

Water-using appliances like humidifiers and pipes can harbor and nurture waterborne bacteria. People may become ill from some of these pathogens. When the water is stagnant (not moving) or has not received enough chemical treatment (such as chlorine), germs survive and thrive in it.

Understanding the source of your tap water and how to utilize it properly for uses other than drinking is crucial.

Public water utilities are required by federal and state legislation to supply drinking water that satisfies specific quality and safety requirements. However, tap water is not sterile; therefore, microorganisms may be present. A tiny number of bacteria that are present in the environment naturally may still be present in the water even when the public water system is operating well. Under certain conditions, these microorganisms can proliferate and develop in the pipes within a house or building. This may occur, for instance, if you don't open the taps for extended periods of time and the water in the pipes remains still.

Most individuals are probably aware that swallowing certain waterborne bacteria can result in gastrointestinal ailments like vomiting or diarrhea. However, these microorganisms can also result in skin, eyes, brain, or lung diseases. When you turn on the water, bacteria from biofilm may emerge from the faucet, showerhead, or other water appliances, particularly if you haven't used the water in a long time (a week or longer). People may become ill from some of these bacteria when the water is inhaled as a mist, makes contact with an exposed wound, goes up the nose (like in the case of using a neti pot), is used to rinse or store contact lenses or is splashed in someone's eyes while they are wearing contacts.

The majority of healthy individuals are not ill when exposed to the bacteria found in pipes. However, some demographics can be more susceptible to infection. Among these groups are: individuals who are fifty years or older; current or former smokers; individuals suffering from long-term lung conditions, such as emphysema or chronic obstructive pulmonary disease (COPD); individuals with medical conditions or those on medications that impair the body's defenses against infection (such as those with compromised immune systems due to HIV, cancer, or organ transplantation); individuals with comorbid conditions including diabetes, liver failure, or kidney failure; babies younger than six months and those who wear contact lenses.

You can take steps to protect yourself from waterborne germs in your home: flush your faucets and showerheads if you haven't used them recently; clean and maintain all devices that use water; communicate with your water utility; and keep private water sources safe.

Testing and Evaluating Water Sources

One of the safest water supplies in the world is found in the United States. The community water systems that supply tap water to more than 90% of Americans are governed by regulations pertaining to safe drinking water. The quality of drinking water varies from location to location based on the state of the source water and the level of treatment applied, but it must adhere to EPA (Environmental Protection Agency) standards. Community water systems adhere to the Safe Drinking Water Act's regulations. The SDWA is at least as protective as the federal guidelines set by the EPA. Several states have their own drinking water regulations that must be followed. Water testing schedules, procedures, and guidelines for drinking water quality are all part of the SDWA regulations.

Although tap water in the United States is regarded as one of the safest globally, contamination can still happen.

Numerous potential causes of contamination exist, such as: sewage releases; minerals and substances that are found in nature (such as uranium, radon, and arsenic); local land use methods (such as concentrated feeding operations, livestock, herbicides, and fertilizers); manufacturing processes (heavy metals, cyanide, etc.); and malfunctioning on-site wastewater treatment systems (for example, septic systems).

Furthermore, situations that raise the danger of contamination can result from drinking water that has not been adequately treated, or that passes through pipelines or other distribution systems that are not properly maintained.

Water system officials may issue a water advisory to help safeguard the public's health when they discover a problem with the drinking water supply (such as contamination). Certain toxins in our water can cause neurological diseases, reproductive troubles, and gastrointestinal illnesses, among other health difficulties. Those with compromised immune systems, the elderly, pregnant women, small children, and infants may be particularly vulnerable to disease.

The EPA establishes guidelines and limits for the existence and concentrations of more than ninety pollutants in public drinking water, including lead, salmonella, E. Coli, and disinfection byproducts.

Each community water supplier is required to give its clients an annual report, which is also occasionally referred to as a Consumer Confidence Report, or "CCR." The study offers details on the quality of drinking water in the area, including where the water comes from, existing pollutants, and how the public may help safeguard the water.

Bottled Water

The Food and Drug Administration (FDA) regulates bottled water safety and bases its standards on the EPA standards for tap water. For the majority of healthy people, water is deemed safe if these requirements are fulfilled. The FDA's good manufacturing procedures for processing and bottling drinking water must be adhered to by the bottled water sector. To find out where your bottled water comes from and how it has been treated to make it safe to drink, read the label. Although bottled water does not yet have a standard label, the label may provide information about the water's treatment method. To find out additional information, look for the firm that bottled the water's toll-free number or website address on the label.

Individuals undergoing chemotherapy, have HIV, diabetes, organ transplants, or other immune system-compromising conditions must exercise extra caution regarding the water they drink. In patients with compromised immune systems, the parasite Cryptosporidium can induce severe or persistent sickness, sometimes even life-threatening symptoms. This parasite is more likely to cause a minor illness in healthy individuals. When purchasing bottled water, look for the following treatments that guard against Cryptosporidium: reverse osmosis, distillation, and filtration with an absolute 1-micron filter.

Even while outbreaks of disease linked to bottled water are rare, they do happen. It is critical that producers, retailers, and end users of bottled water take the following actions: safeguard and treat water appropriately prior to bottling; uphold quality control procedures during production; safeguard bottled water during transportation and storage; and guard against contamination following consumer purchase. Your health can be negatively impacted by contaminated bottled water, as it can lead to neurological abnormalities, reproductive issues, and gastrointestinal illnesses. Certain pollutants may be more prone to cause illness in newborns, young children, pregnant women, the elderly, and those with compromised immune systems.

Filtering and Purifying Water at Home

Water filters work differently. Some can remove hazardous chemicals or bacteria from your water, while others can improve the flavor of your water. Not everyone needs a water filter, and no single filter can remove every kind of pollution from drinking water. In actuality, the water that flows from your tap has trace amounts of numerous additional chemicals. Some of these are good for you, like fluoride, which helps prevent tooth decay, and the right amount of a disinfectant, like chlorine, which helps keep your water free of bacteria. Other potentially dangerous materials found in water include lead and the bacteria Cryptosporidium. Both healthy and toxic elements can be eliminated from your water using filters. Filtering your water may or may not be necessary, depending on your situation.

It might be difficult to determine whether you need a filter and the ideal one for you because there are so many different kinds available. Filtration is one way to safeguard your health when drinking water that is unsafe in many parts of the world. When deciding whether or not to use a water filter, and which kind and features best suit your unique requirements and tastes, take into account the following measures.

An independent group called NSF International creates standards for products' public health. Searching for an NSF certification on the label is one approach to learning what a water filter accomplishes. To find out what a product is certified to safeguard against, search for it online in the NSF database. The NSF standards 41 (taste and odor), 53 (cyst reduction), 58 (reverse osmosis), and 62 (distillation) are a few that apply to water treatments.

The size of the microscopic holes in a filter that allow water to pass through is known as the pore size. For example, all of the pores in a filter that has an "absolute" pore size of 1 micron are 1 micron or smaller. This implies that any impurity, such as Cryptosporidium, larger than one micron will be captured by the filter and excluded from the filtered water. The average pore size of filters with a "nominal" or "mean" pore size of 1 micron is 1 micron, but there are some pores that are smaller and others that are larger than 1 micron. As a result, impurities like Cryptosporidium might pass through the larger pores and enter the water that you are filtering (CDC, 2023).

Step 1: Know Your Water Source

Your specific water supply's quality, taste and odor preferences, and any unique health issues in your home will determine how much water filtration you require. It is beneficial to become more knowledgeable about your water source before selecting the ideal filter for your household's needs. Determining whether you want or need a filtration system and which kind would best suit your needs requires first determining what contaminants might already be in your water.

If you use a private well for your water, you should have it tested at least once a year by a lab with state certification. Make sure the person you choose will test for typical contaminants in your area, such as total coliform bacteria, nitrates, total dissolved solids, pH levels, and others. For information on toxins in your area and recommendations on water tests you can order, get in touch with your local health department. Additionally, if a child moves in or someone in your home gets pregnant, you should have your water tested. If the color, taste, or odor of your water changes, you should consider testing it. You can find out the amounts of various toxins in your water by having regular testing done at an accredited lab. You can choose a filter that can lessen toxins in your water once you are aware if any of them are present in excess of what is advised.

A yearly report on your water will be sent to you if your water comes from a public system. All public water systems are required by the EPA to provide an annual report regarding the quality of drinking water and the levels of contaminants to individuals whose water comes from public sources.

Step 2: Think About Why You Are Considering a Filter

These are a few common reasons for using water filters. The first step in selecting the ideal water treatment system for you is determining what you require or desire from it.

Taste

It may surprise you to hear that the most popular pitcher and refrigerator filters only partially filter out bacteria and other impurities; their primary purpose is to make your water taste better. The taste of tap water bothers some people. This can occasionally be attributed to disinfectants (such as chlorine) that help keep the water free of bacteria. The flavor of water can occasionally be altered by minerals or other harmless naturally occurring pollutants like molecules that include sulfur.

Unpleasant tastes and smells can be decreased with the use of activated carbon filters, which are the kinds of filters used in many refrigerators and pitcher filters. Reverse osmosis systems can lower the concentrations of common pollutants like lead while also enhancing taste. Verify the label to make sure the specific filter you are thinking about addresses taste and odor (NSF 42). Remember that using a whole-house filter designed to remove chlorine from the air may actually increase the number of microorganisms growing in your plumbing.

Most of the dangerous pollutants are taste, smell, or sight-invisible. Certain dangerous pollutants, like gasoline or other chemicals, can give water a terrible taste or smell. One such contamination is volatile organic compounds (VOCs), which can occasionally pollute private wells. VOCs come in a wide variety of forms, and each one has a unique set of health impacts. Some harm mucous membranes, cause cancer, irritate the skin, or harm the kidneys, liver, or nervous system. It might be necessary to conduct additional tests to find out which VOCs are in your water in order to choose the optimal filter. For VOCs, it is recommended to use a whole-house filter system or a point-of-entry filter system (where your water pipe enters your home). These systems offer safe water for drinking, cooking, and bathing. Activated carbon filters can remove some VOCs.

Consider having your well water analyzed if you have a private well and you've noticed a change in the flavor of your water. Notify your water provider if you have a public water system and the flavor of your water has changed. Just remember that water's taste and smell—or lack thereof—does not always indicate how pure it is of chemicals and bacteria.

Lead

Homes with lead service lines connecting them to the main water line may contain lead in their drinking water. Lead-soldered plumbing fixtures such as galvanized iron pipes, brass or chrome-plated brass faucets, and other fixtures may still be found in homes without lead service lines. Lead-containing plumbing materials can undergo a chemical reaction that releases lead into drinking water. The dissolution or wearing away of metal from pipes and fittings is referred to as corrosion.

Finding any possible lead sources in your home plumbing and service line is the best way to determine the level of your drinking water exposure. It is recommended to control or eliminate all sources of lead exposure for children, as there is currently no known safe blood level for young children. Because lead can be hazardous to human health even at low exposure levels, the EPA has set a zero maximum lead contamination level goal in drinking water.

You can take steps to lower the level of lead in your drinking water and lessen your risk of exposure if you are worried about lead in water or know that lead is present in your plumbing. By drinking or using only tap water that has been passed through a "point-of-use" filter approved by an independent testing organization to reduce or eliminate lead (NSF/ANSI standard 42 for particulate removal and NSF/ANSI standard 53 for lead removal), you can minimize or completely eliminate your exposure to lead in tap water. Use a filter on any drinking or cooking water if your home has a lead service line.

To lessen the possibility of being exposed to lead from lead plumbing in your home, you can flush your water. When the water has been off and lying in the pipes for longer than six hours, this is very crucial. Make sure your home's pipes are clear before drinking by running the faucet, taking a shower, doing the laundry, or doing a load of dishes. Whether or whether your home has a lead service line and how long it is will determine how long you need to run the water. Use only cold water straight from the faucet for cooking or drinking.

Lead concentrations can be higher in warm or hot water that is just out of the faucet (CDC, 2023).

Arsenic

One heavy metal that is frequently discovered in groundwater sources is arsenic, which can be found in both private wells and public water systems that use groundwater. Arsenic can cause cancer and is linked to various health issues. There are two possible types of arsenic: pentavalent and trivalent. To choose the right filter, it is critical to understand the type (or "species") of arsenic in your water. Pentavalent arsenic is more common in water treated with chlorine; it can be eliminated using filters with the NSF standard 53 or 58 mark. In the event that your water is untreated, you may need to undergo further treatment (also known as a "pre-oxidation step") in order to change trivalent arsenic to pentavalent arsenic before the water is filtered. Distillation is a very efficient method of eliminating arsenic from water, but because it requires more energy and takes longer than other methods, it is not as useful for residential usage (CDC, 2023).

Nitrates

Chemicals known as nitrates enter groundwater through sewage, manure, fertilizer, septic systems, and natural deposit erosion. The oxygen-carrying ability of your red blood cells is hampered by nitrates. For certain people and infants, this may be harmful. If the water you use comes from a public water supply, nitrate levels are kept under observation and managed. If you have a private well, you should test the water to see if nitrates are present. You have two options if testing shows that your water contains large amounts of nitrates:

reverse osmosis (NSF 58) or distillation (NSF 62). Nitrates cannot be eliminated by boiling or filtration (CDC, 2023).

Compromised Immune Systems

Individuals who are immunocompromised or whose immune systems have been weakened by organ transplants, AIDS, chemotherapy, or other medical conditions should speak with their healthcare provider about whether they should install a water treatment system to make sure their water is free of bacteria, particularly Cryptosporidium. Both filters with a pore size of "absolute 1 micron" and those labeled "reverse osmosis" offer protection against Cryptosporidium. Additionally, you can check the label for the NSF 53 or NSF 58 standards. Cryptosporidium is particularly concerning for those with weakened immune systems.

Filters intended to eliminate Crypto (if a package label contains any of the following four messages, the filter should be capable of eliminating Crypto):

- reverse osmosis (NSF testing included or not)
- absolute pore size (with or without NSF testing) of one micron or less
- NSF Standard 53 or NSF Standard 58 tested and approved for the elimination of cysts
- NSF Standard 53 or NSF Standard 58 tested and approved for cyst reduction

Furthermore, immunocompromised individuals should not change their own water filters because doing so could expose them to the impurities the filter collects and raise their risk of infection. When used correctly, certain filters can get rid of parasites like Giardia and Cryptosporidium. But they might not get rid of bacteria and viruses. It's possible that your filtered water still contains germs (CDC, 2023).

Nasal Rinsing

Use safe water to reduce your chance of contracting Naegleria fowleri if you are making a solution to irrigate, flush, or rinse your sinuses (for instance, using a neti pot, sinus rinse bottle, or other irrigation device) or putting water in your nose as part of a religious practice. This tiny ameba travels up the nose to the brain, where it causes a rare illness that can be fatal.

To reduce your chance of contracting the infection, do at least one of the following:

- Boil: Use water that has been boiled and then allowed to cool. Boil for three minutes for heights higher than 6,500 feet.
- Filter: Make use of a filter to get rid of some bacteria that like water. Look for "NSF 53" or "NSF 58" on the label, and filter labels that state, "absolute pore size of 1 micron or smaller."
- Buy: Use water with a label specifying that it contains distilled or sterile water.
- Disinfect: To make sure your water is free of Naegleria, find out how to disinfect it. When applied

correctly and at the appropriate time, chlorine bleach functions as a disinfectant against this pathogen.

Hard Water

Hard water, or water with high mineral content, such as calcium and magnesium, can cause scaly buildup and hinder soap lathering. This issue can be resolved with water softeners. Since water softeners employ ion exchange technology, they cannot be considered filters and cannot shield you from bacteria found in the water. Beneficial minerals are also taken out of the water by water softeners.

Step 3: Consider How the Filter Fits Your Home, Lifestyle, and Budget

Customers can choose from a wide variety of filters. It's critical to understand what filters do and do not do because no filter can completely remove all pollutants.

Check the label to find out if it has NSF certification. If so, you can find out more about the threats that a specific model is certified to shield you from by searching the NSF database. The contaminants that are decreased are usually listed on the labels of water filters, which might aid in decision-making. Since not all salespeople will be conversant with your needs, make sure you carefully read labels and double-check the manufacturer's claims using an impartial source.

Remember that most manufacturers have a wide variety of filters. Salespeople may be able to assist you in choosing the right product but keep in mind that they are occasionally compensated to promote a specific brand. Before making a purchase, verify that the filter will serve your needs by reading the fine print and verifying the claims made on the packaging.

Never assume that just because a filter gets rid of one contaminant, it gets rid of others, too. Chemical-removing filters frequently fail to remove germs as well, and vice versa. Fluoride may be eliminated by some water treatment technologies, such as distillation systems, ion exchange, and reverse osmosis. Fluoride supplements may be necessary for kids whose water has less than 0.6 parts per million of fluoride. For particular advice, speak with your child's pediatrician or dentist (CDC, 2023).

The cost of various filtration systems varies greatly; basic systems can be purchased for less than $20, while complicated systems can reach several hundred dollars and require expert installation. In addition to the initial cost of purchase and installation, also take into account the system's cost, schedule, and ease of maintenance, including filter cartridge changes. Every water treatment system needs maintenance to keep functioning correctly.

While some filters take a long time to filter a lot of water, others do so rapidly. You might not require a quick filter if the filter is solely used for your personal drinking water. Water filter pitchers, end-of-tap or faucet-mounted filters, faucet-integrated (built-in) filters, on-counter filters, under-sink filters, and whole-house treatment units are examples of filters that are frequently seen in homes and stores.

Water Filter Pitchers

Pitchers with built-in filters that water must pass through before being emptied out for drinking or other purposes are known as water filter pitchers. These pitchers are filled from the top.

Pros: Inexpensive to purchase, no installation, easy to use

Cons: Slow filtering; varies by model and pore size; filters need to be changed frequently.

Refrigerator Filters

A built-in filter in many refrigerators delivers water via the door and powers an ice maker that operates automatically.

Pros: Often included with refrigerators, frequently enhance the flavor of water, may filter water used to make ice, and are simple to use

Cons: Regular filter replacement is required

Faucet-Mounted Filters

Filtration devices that are attached to normal faucets can be turned on and off to switch between filtered and unfiltered water flow.

Pros: Reasonably priced, simply switchable between filtered and unfiltered water

Cons: Not compatible with all faucets; slows the flow of water

Faucet-Integrated (Built-In) Filters

Installing faucet-integrated filtration systems is necessary since these faucets are made with built-in filters rather than a connected filter, as with faucet-mounted systems.

Pros: Changing from filtered to unfiltered water is simple. Offers a sleek appearance.

Cons: Installation is often necessary and costly.

On-Counter Filters

These filters sit beside the sink, connected to your faucet and filter water through a separate countertop dispenser.

Pros: High-capacity filtration without altering your plumbing.

Cons: Requires countertop space and may not suit smaller kitchens.

Under-Sink Filter

Installed beneath a sink, under-sink filtration systems route water through a pipe to a faucet that is specifically designed for the filter.

Pros: Filter a lot of water and don't take up much countertop space.

Cons: Frequently costly, may call for plumbing adjustments

Whole-House Water Treatment

Water treatment systems for the entire house purify all incoming water, not only what is intended for human consumption.

Pros: All water entering your home is treated, which is beneficial if you have hard water or volatile organic compounds (VOCs).

Cons: Frequently costly, may need plumbing adjustments, expert maintenance, and chlorine-removing filters may promote the growth of bacteria in all of your home's pipes.

Step 4: Maintain Your Filters

It's important to follow all manufacturer maintenance requirements because poorly maintained filters might cause more harm. Regular filter changes will maintain the quality of your drinking water. Regular regeneration, also known as flushing, is required for water softeners. It is recommended that someone without immunocompromised status replace the filter cartridges since filters gather microorganisms from water. It is recommended to wear gloves while changing the cartridges. After changing the cartridges, everyone involved should wash their hands even if they had the gloves on.

Equipment Basics

The goal of the purification process is to make water safer and cleaner for human use. A consumer's health might often be in danger from the presence of germs, pollutants, and other impurities in water. These contaminants are eliminated throughout the water purification process, leaving behind clean, safe-to-drink water. Different kinds of water contamination can be treated with different kinds of water purifying techniques. For instance, a procedure known as bacteria purification can be used to purify water that has been tainted with dangerous germs. Chemical purification is used to clean water contaminated with chemicals and minerals, whereas viral purification is used to clean water contaminated with viruses. There are various systems that you can get to purify your drinking water.

Water Filter

One of the most popular devices for water purification is a water filter. A water filter is an apparatus designed to physically filter pollutants out of water. There are numerous kinds of water filters that work well for various kinds of contaminants. For instance, contaminants like grit and sand are removed from water using a sediment filter, whereas bacteria can be removed using a carbon filter. A distillation filter can be used to remove chemicals and bacteria, whereas a reverse osmosis filter is designed to remove minerals.

Chemical Purification Systems

These water purifying systems work well against chemical contaminants, but they are unable to handle contaminants caused by bacteria or viruses. Minerals and some chemical contaminants, such as chlorine, are frequently present in water. You can get clean water by breaking down the chemical contaminants in the water by the chemical purification system.

Ultraviolet Sterilization

Ultraviolet (UV) sterilization uses UV radiation to destroy viral contaminants in water. UV sterilization is frequently combined with other techniques for purifying water. Low mineral content water can be effectively sterilized by UV light because excessive mineral content can obstruct the light from the UV sterilizing apparatus. Water contaminated by bacteria, viruses, parasites, or other animal or human pathogens can be cleaned up using UV sterilization. Since UV sterilization is a chemical-free method of water filtration, it is an excellent option for low-mineral water. Yellow fever, dengue fever, hepatitis, and stomach flu are just a few of the viruses that can be effectively combated by UV sterilization.

Minimum Equipment Treatment

In many places of the world, communities lack the ability to build piped water systems that serve the entire community. More than 2 billion people do not have access to drinking water that is properly managed. Water that is treated at home or at a community's point of use or collection—like schools and community centers—is referred to as household water treatment. This kind of treatment can minimize diarrheal and other waterborne infections and make water safe to use in the absence of a piped water supply.

When selecting water treatment techniques for a home or community, there are a number of things to take into account, including the quality of the water, the ability to install WASH facilities, cultural acceptability, accessibility, technology availability, consistent and long-term use, and other local conditions. There are numerous techniques to rid water of dangerous bacteria, namely boiling, chlorination, flocculant disinfection, solar disinfection, and slow sand filtration (CDC, 2023).

Boiling water is the most popular and efficient way to destroy bacteria, viruses, and parasites that cause illness. For one minute, bring clear water to a rolling boil (three minutes if the elevation is higher than 6,500 feet). Let the water cool down once it has boiled before using. Store the boiled water in clean, airtight containers.

Chemical disinfection is another widely used technique to ensure the safety of water. Chlorination is a widely used chemical disinfection method that involves killing bacteria and viruses in water by adding a product containing chlorine (such as household bleach, calcium hypochlorite, or sodium hypochlorite). Chlorine dioxide and iodine are two other chemical disinfectants that can be used to disinfect water effectively. Small levels of chlorine, iodine, or chlorine dioxide in water do not have a negative impact on health and can help prevent the spread of waterborne illnesses. Most hazardous or disease-causing viruses and bacteria can be killed by items containing chlorine, but boiling is a more effective way of destroying highly resistant pathogens, like the parasites Giardia and Cryptosporidium. If you properly follow the manufacturer's directions, Cryptosporidium can be killed with chlorine dioxide tablets.

A substance is added to water during the flocculation-disinfection process, which allows particles to aggregate into bigger clusters that can be extracted from the water. A popular powder flocculant-disinfectant is PUR sachets. The PUR product has calcium hypochlorite, a disinfectant, and powdered ferric sulfate, a flocculant. After adding the contents of the PUR sachet to ten liters of water, swirl for five minutes. The water is poured through a cloth filter into a secondary water container for safe storage and use once the solids have settled to the bottom of the container from swirling. PUR also contains chlorine for residual protection. The best conditions for using this procedure are turbid (cloudy) water.

Through a variety of biological and physical mechanisms, slow sand filtration efficiently eliminates microorganisms and turbidity (cloudiness) in a single treatment step. Layers of components are stacked vertically to form a slow sand filter. When built, the filter has a gravel layer to support the fine sand bed, a tank, a network of underdrains to collect the filtered water, and a flow regulator to adjust the filtration rate. This filtration procedure doesn't require adding any chemicals.

Chapter 5:

Minimum Requirements for Quality Water

Drinking water. It rushes through invisible pipes in the walls, under the streets, and all over the world to fill tens of billions of glasses, cups, and bottles every day and to sate thirst—the most basic human need. For millions of years, one of the most crucial pieces of information our ancestors possessed was firsthand knowledge about the source of our water. That intimacy is now gone. As if by magic, water flows when we turn on a tap. We now acknowledge the illusion to be true. Most of us don't really know where our drinking water comes from. We assume that it will always be there. We assume that it will always be safe.

There are many assumptions on the path to catastrophe. The greatest waterborne pandemic in American history occurred in 1993, not centuries earlier. Not only does waterborne illness still exist, but its frequency is unknown. In most cases, drinking water is never even identified as the source of a waterborne disease since our mechanism for doing so is inadequate. Research indicates that millions of Americans may become ill each year from drinking contaminated water. In a large portion of the underdeveloped world, waterborne illness is commonplace. Unsafe drinking water is killing people continuously; this week alone, about forty thousand people will perish from it.

In less than a month in 1994, cholera killed 60,000 people in a packed Goma, Democratic Republic of the Congo, refugee camp. In human history, it was the worst waterborne disease outbreak. For many of us, the tragedy of Goma is so outside of our experience that it seems distant and abstract. Although there appears to be a huge gap between the sanitary comforts of the developed world and an epidemic in Goma, there may not be as much of a gap as we imagine. Waterborne typhoid fever was one of the main causes of death in the US a century ago.

The systems we have created to transport and clean our drinking water and waste are the only things separating us from Goma. Most of us are unaware of how our water supplies function. Invisibility breeds complacency. Although we now consider these systems to be foolproof, many of our water treatment facilities have been in use since the early 1900s, and the technology we use to treat the majority of our drinking water is nearly a century old.

Treated sewage makes up at least a portion of the water coming from these old plants. Our water treatment plants' intakes frequently receive farm runoff, industrial waste, and sewage—both treated and untreated. Research has indicated that certain pathogens or germs that cause disease can and do find their way into drinking water supplies. These pathogens can occasionally cause devastating outbreaks and frequently cause isolated episodes of illness. Although cholera is more fatal than these illnesses, this may not always be the case.

We are surrounded by illness. It may not be water-borne illnesses like cholera or dysentery, but cancer and heart disease are rife. Our environments, including the water we drink, contribute to all these diseases that are prevalent today. In this chapter, we explore the different contaminants found in our water, how many of these contaminants can be found in our water, and the water treatment options available to us.

Contaminants to Be Aware of in Water

Any material or substance that is physical, chemical, biological, or radioactive in water is considered a "contaminant" under the Safe Drinking Water Act. Thus, anything other than water molecules is included in the exceedingly broad legal definition of "contaminant." It is reasonable to assume that some contaminants will be present in drinking water, even only in trace amounts. While some contaminants in drinking water may be safe to consume in certain amounts, others may be dangerous. Contaminants alone do not always mean that drinking the water is dangerous for your health.

Examples of each of the major categories of contaminants found in drinking water are shown below:

- Physical impurities mostly affect the way water looks or has other physical characteristics. Sediment and organic matter carried by soil erosion into lakes, rivers, and streams are examples of physical pollutants.

- Compounds or elements make up chemical pollutants. These pollutants could be man-made or naturally occurring. Chemical pollutants include things like nitrogen, bleach, salts, metals, insecticides, bacteria-produced toxins, and medications used on humans or animals.

- Waterborne biological contaminants are living things. They are also known as microorganisms or microbiological pollutants as well. Contaminants classified as biological or microbial include bacteria, viruses, parasites, and protozoa.

- Chemical elements that have an uneven proton and neutron count are known as radiological pollutants because they produce unstable atoms that are capable of releasing ionizing radiation. Radiation-related pollutants include uranium, plutonium, and cesium.

Different substances, bacteria, and radionuclides can contaminate surface waters and aquifers. In the United States, the prevalence of waterborne illnesses like cholera, hepatitis, and typhoid has significantly decreased as a result of drinking water disinfection. Various methods may be employed to purify drinking water, depending on the attributes and impurities present in the water supply.

The common sources of contaminants found in our drinking water include agriculture and industry, human and animal waste, water treatment and distribution, and natural sources.

Aquifers may get contaminated by organic solvents, petroleum products, and heavy metals that leak from storage or disposal sites. Fertilizers and pesticides can seep into aquifers or be carried into lakes and streams by snowmelt and rainfall runoff.

Hazardous bacteria can enter drinking water sources through human waste from sewage and septic systems, animal feedlot waste, and wildlife waste. E. Coli, Giardia, and Cryptosporidium are major pollutants.

Many pollutants can be removed during treatment, but they can also leave behind byproducts (such as trihalomethanes) that could be dangerous. Additionally, water can get contaminated once it reaches the distribution system due to plumbing materials consisting of lead or copper corroding or from a breach in the pipe system.

Because of the high concentrations of certain toxins in the subsurface conditions, some groundwater is unsafe to drink. For example, naturally occurring arsenic, other heavy metals, or radionuclides can be absorbed by groundwater as it passes through rock and soil.

Some of the germs that can contaminate tap water are: cryptosporidium, legionella, campylobacter, norovirus, E. coli O157, rotavirus, enterovirus, salmonella, giardia, shigella, and hepatitis A virus. On the chemical side, contaminants include arsenic, copper, lead, nitrate, and radon.

Effect on Health

Unsafe contaminant levels in drinking water can have negative impacts on the brain system, the reproductive system, the gastrointestinal tract, and chronic disorders like cancer. The type of contamination, its concentration in the water, an individual's susceptibility, the quantity of water ingested, and the length of exposure are some variables that can affect whether a contaminant will have an adverse effect on health.

Drinking water exposure to chemicals can have a number of short- and long-term health impacts. Chemical exposure at high concentrations can cause skin discoloration or more serious issues like harm to the nervous system or other organs, impacts on development or reproduction, etc. Long-term, chronic illnesses like cancer can result from prolonged exposure to smaller dosages. There is still much to learn about the impact of various pollutants in drinking water.

The majority of potentially fatal microbe-caused aquatic illnesses, such as cholera and typhoid fever, are uncommon in the US today. The more prevalent infections brought on by bacteria, viruses, and parasites can cause kidney failure, vomiting, diarrhea, headaches, and stomach pain. Additionally, infectious disorders like hepatitis can manifest. In individuals with significantly damaged immune systems (such as cancer and AIDS patients), hepatitis can sometimes be fatal. It can also be severe in persons with reduced immune systems, such as infants and the elderly.

Safe Levels of Various Substances in Water

Below (on next page) is a table with all the information on the different contaminants present in water and the maximum allowable traces of the contaminants in our drinking water (CDC, 2023).

Contaminant	MCLG	MCL or TT	Potential Health Effects from Long-Term Exposure Above the MCL (unless specified as short-term)	Sources of Contaminant in Drinking Water
Cryptosporidium	zero	TT	Gastrointestinal illness (such as diarrhea, vomiting, and cramps)	Human and animal fecal waste
Giardia lamblia	zero	TT	Gastrointestinal illness (such as diarrhea, vomiting, and cramps)	Human and animal fecal waste
Heterotrophic plate count (HPC)	n/a	TT	HPC has no health effects; it is an analytic method used to measure the variety of bacteria that are common in water. The lower the concentration of bacteria in drinking water, the better maintained the water system is.	HPC measures a range of bacteria that are naturally present in the environment
Legionella	zero	TT	Legionnaire's Disease, a type of pneumonia	Found naturally in water; multiplies in heating systems
Total Coliforms (including fecal coliform and *E. Coli*)	zero	5.0%	Not a health threat in itself; it is used to indicate whether other potentially harmful bacteria may be present	Coliforms are naturally present in the environment, as well as feces; fecal coliforms and *E. coli* only come from human and animal fecal waste.
Turbidity	n/a	TT	Turbidity is a measure of the cloudiness of water. It is used to indicate water quality and filtration effectiveness (such as whether disease-causing organisms are present). Higher turbidity levels are often associated with higher levels of disease-causing microorganisms such as viruses, parasites, and some bacteria. These organisms can cause symptoms such as nausea, cramps, diarrhea, and associated headaches.	Soil runoff
Viruses (enteric)	zero	TT	Gastrointestinal illness (such as diarrhea, vomiting, and cramps)	Human and animal fecal waste

Bromate	zero	0.010	Increased risk of cancer	Byproduct of drinking electrical, aerospace, and defense industries
Cadmium	0.005	0.005	Kidney damage	Corrosion of galvanized pipes; erosion of natural deposits; discharge from metal refineries; runoff from waste batteries and paints
Chromium (total)	0.1	0.1	Allergic dermatitis	Discharge from steel and pulp mills; erosion of natural deposits
Copper	1.3	TT; Action Level=1.3	Short-term exposure: Gastrointestinal distress Long-term exposure: Liver or kidney damage People with Wilson's Disease should consult their personal doctor if the amount of copper in their water exceeds the action level	Corrosion of household plumbing systems; erosion of natural deposits
Cyanide (as free cyanide)	0.2	0.2	Nerve damage or thyroid problems	Discharge from steel/metal factories; discharge from plastic and fertilizer factories
Fluoride	4.0	4.0	Bone disease (pain and tenderness of the bones). Children may get mottled teeth	Water additive which promotes strong teeth; erosion of natural deposits; discharge from fertilizer and aluminum factories
Lead	zero	TT ; Action Level=0.015	Infants and children: Delays in physical or mental development; children could show slight deficits in attention span and learning abilities Adults: Kidney problems; high blood pressure	Corrosion of household plumbing systems; erosion of natural deposits
Mercury (inorganic)	0.002	0.002	Kidney damage	Erosion of natural deposits; discharge from refineries and factories; discharge from

				runoff from landfills and croplands
Nitrate (measured as Nitrogen)	10	10	Infants below the age of six months who drink water containing nitrate in excess of the MCL could become seriously ill and, if untreated, may die. Symptoms include shortness of breath and blue-baby syndrome.	Runoff from fertilizer use; leaking from septic tanks, sewage; erosion of natural deposits
Nitrite (measured as Nitrogen)	1	1	Infants below the age of six months who drink water containing nitrite in excess of the MCL could become seriously ill and, if untreated, may die. Symptoms include shortness of breath and blue-baby syndrome.	Runoff from fertilizer use; leaking from septic tanks, sewage; erosion of natural deposits
Selenium	0.05	0.05	Hair or fingernail loss; numbness in fingers or toes; circulatory problems	Discharge from petroleum refineries; erosion of natural deposits; discharge from mines
Thallium	0.0005	0.002	Hair loss; changes in blood; kidney, intestine, or liver problems	Leaching from ore-processing sites; discharge from electronics, glass, and drug factories

Contaminant	MCLG (mg/L)	MCL or TT (mg/L)	Potential Health Effects from Long-Term Exposure Above the MCL (unless specified as short-term)	Sources of Contaminant in Drinking Water
Acrylamide	zero	TT	Nervous system or blood problems; increased risk of cancer	Added to water during sewage/wastewater treatment
Alachlor	zero	0.002	Eye, liver, kidney, or spleen problems; anemia; increased risk of cancer	Runoff from herbicide used on row crops
Atrazine	0.003	0.003	Cardiovascular system or reproductive problems	Runoff from herbicide used on row crops
Benzene	zero	0.005	Anemia; decrease in blood platelets; increased risk of cancer	Discharge from factories; leaching from gas storage tanks and landfills
Benzo(a)pyrene (PAHs)	zero	0.0002	Reproductive difficulties; increased risk of cancer	Leaching from linings of water storage tanks and distribution lines
Carbofuran	0.04	0.04	Problems with blood, nervous system, or reproductive system	Leaching of soil fumigant used on rice and alfalfa
Carbon tetrachloride	zero	0.005	Liver problems; increased risk of cancer	Discharge from chemical plants and other industrial activities
Chlordane	zero	0.002	Liver or nervous system problems; increased risk of cancer	Residue of banned termiticide
Chlorobenzene	0.1	0.1	Liver or kidney problems	Discharge from chemical and agricultural chemical factories
2,4-D	0.07	0.07	Kidney, liver, or adrenal gland problems	Runoff from herbicide used on row crops

Dalapon	0.2	0.2	Minor kidney changes	Runoff from herbicide used on rights of way
1,2-Dibromo-3-chloropropane (DBCP)	zero	0.0002	Reproductive difficulties; increased risk of cancer	Runoff/leaching from soil fumigant used on soybeans, cotton, pineapples, and orchards
o-Dichlorobenzene	0.6	0.6	Liver, kidney, or circulatory system problems	Discharge from industrial chemical factories
p-Dichlorobenzene	0.075	0.075	Anemia; liver, kidney, or spleen damage; changes in blood	Discharge from industrial chemical factories
1,2-Dichloroethane	zero	0.005	Increased risk of cancer	Discharge from industrial chemical factories
1,1-Dichloroethylene	0.007	0.007	Liver problems	Discharge from industrial chemical factories
cis-1,2-Dichloroethylene	0.07	0.07	Liver problems	Discharge from industrial chemical factories
trans-1,2-Dichloroethylene	0.1	0.1	Liver problems	Discharge from industrial chemical factories
Dichloromethane	zero	0.005	Liver problems; increased risk of cancer	Discharge from drug and chemical factories
1,2-Dichloropropane	zero	0.005	Increased risk of cancer	Discharge from industrial chemical factories
Di(2-ethylhexyl) adipate	0.4	0.4	Weight loss, liver problems, or possible reproductive difficulties.	Discharge from chemical factories
Di(2-ethylhexyl) phthalate	zero	0.006	Reproductive difficulties; liver problems; increased risk of cancer	Discharge from rubber and chemical factories
Dinoseb	0.007	0.007	Reproductive difficulties	Runoff from herbicides used on soybeans and vegetables
Dioxin (2,3,7,8-TCDD)	zero	0.00000003	Reproductive difficulties; increased risk of cancer	Emissions from waste incineration and other

				combustion; discharge from chemical factories
Diquat	0.02	0.02	Cataracts	Runoff from herbicide use
Endothall	0.1	0.1	Stomach and intestinal problems	Runoff from herbicide use
Endrin	0.002	0.002	Liver problems	Residue of banned insecticide
Epichlorohydrin	zero	TT	Increased cancer risk and, over a long period of time, stomach problems	Discharge from industrial chemical factories; an impurity of some water treatment chemicals
Ethylbenzene	0.7	0.7	Liver or kidney problems	Discharge from petroleum refineries
Ethylene dibromide	zero	0.00005	Problems with the liver, stomach, reproductive system, or kidneys; increased risk of cancer	Discharge from petroleum refineries
Glyphosate	0.7	0.7	Kidney problems; reproductive difficulties	Runoff from herbicide use
Heptachlor	zero	0.0004	Liver damage; increased risk of cancer	Residue of banned termiticide
Heptachlor epoxide	zero	0.0002	Liver damage; increased risk of cancer	Breakdown of heptachlor
Hexachlorobenzene	zero	0.001	Liver or kidney problems; reproductive difficulties; increased risk of cancer	Discharge from metal refineries and agricultural chemical factories
Hexachlorocyclopentadiene	0.05	0.05	Kidney or stomach problems	Discharge from chemical factories
Lindane	0.0002	0.0002	Liver or kidney problems	Runoff/leaching from insecticide used on cattle, lumber, gardens
Methoxychlor	0.04	0.04	Reproductive difficulties	Runoff/leaching from insecticide used on fruits, vegetables, alfalfa, livestock

Oxamyl (Vydate)	0.2	0.2	Slight nervous system effects	Runoff/leaching from insecticide used on apples, potatoes, and tomatoes
Polychlorinated biphenyls (PCBs)	zero	0.0005	Skin changes; thymus gland problems; immune deficiencies; reproductive or nervous system difficulties; increased risk of cancer	Runoff from landfills; discharge of waste chemicals
Pentachlorophenol	zero	0.001	Liver or kidney problems; increased cancer risk	Discharge from wood-preserving factories
Picloram	0.5	0.5	Liver problems	Herbicide runoff
Simazine	0.004	0.004	Problems with blood	Herbicide runoff
Styrene	0.1	0.1	Liver, kidney, or circulatory system problems	Discharge from rubber and plastic factories; leaching from landfills
Tetrachloroethylene	zero	0.005	Liver problems; increased risk of cancer	Discharge from factories and dry cleaners
Toluene	1	1	Nervous system, kidney, or liver problems	Discharge from petroleum factories
Toxaphene	zero	0.003	Kidney, liver, or thyroid problems; increased risk of cancer	Runoff/leaching from insecticide used on cotton and cattle
2,4,5-TP (Silvex)	0.05	0.05	Liver problems	Residue of banned herbicide
1,2,4-Trichlorobenzene	0.07	0.07	Changes in adrenal glands	Discharge from textile finishing factories
1,1,1-Trichloroethane	0.20	0.2	Liver, nervous system, or circulatory problems	Discharge from metal degreasing sites and other factories

Definitions

The maximum level of a contaminant in drinking water below which there is no known or anticipated risk to health is known as the Maximum Contaminant Level Goal or MCLG. A margin of safety is permitted by MCLGs, which are voluntary public health objectives.

The maximum amount of a pollutant that is permitted in drinking water is known as the Maximum pollutant Level, or MCL. Using the greatest available treatment technology and accounting for cost, MCLs are set as near to MCLGs as is practical. MCLs are legally binding requirements.

The amount of drinking water disinfectant below which there is no known or anticipated risk to health is known as the Maximum Residual Disinfectant Level Goal, or MRDLG. The advantages of using disinfectants to control microbiological pollutants are not reflected in MRDLGs.

Treatment Technique (TT): A necessary procedure used to lower the pollutant level in drinking water.

The maximum amount of disinfection permitted in drinking water is known as the Maximum Residual Disinfection Level, or MRDL. There is strong evidence to support the need to add a disinfectant in order to control microbiological contamination.

Unless otherwise stated, all measurements are in milligrams per liter or mg/L. Parts per million (PPM) are equal to milligrams per liter.

Water Treatment Options for Improving Quality

Apart from the different filters mentioned in the previous chapter, there are other units that you can buy to treat your water (CDC, 2023).

Activated Carbon Filter (Includes Mixed Media That Remove Heavy Metals)

- They absorb organic pollutants that lead to issues with flavor and odor.
- Some designs eliminate byproducts of chlorination.
- Some kinds eliminate pesticides and cleaning agents.
- They are effective in eliminating metals like copper and lead.
- They do not get rid of bacteria, nitrates, or dissolved minerals.

Ion Exchange Unit (With Activated Alumina)

- They eliminate the minerals that give water its "hard" taste, especially calcium and magnesium.
- Certain designs eliminate barium and radium.
- They eliminate fluoride.
- The ion-exchange resin will become coated or clogged and lose its softening function if the water contains iron bacteria or oxidized iron.

Reverse Osmosis Unit (With Carbon)

- They eliminate sodium, nitrates, and other dissolved organic and inorganic substances.
- They eliminate unpleasant odors, tastes, or colors.
- They may also lessen the concentration of certain petrochemicals, dioxins, chloroform, and pesticides.
- They do not completely eliminate all organic and inorganic pollutants.

Distillation Unit

- They eliminate most organic components, heavy metals, radionuclides, hardness, microorganisms, salt, nitrates, and dissolved solids.
- They eliminate bacteria.
- They do not eliminate some pesticides, volatile solvents, and some volatile organic pollutants.
- During times when the cooling coils are not in use, bacteria may repopulate.

Ultraviolet Treatment Systems

- These are treatment processes that use ultraviolet light to disinfect water or reduce the amount of bacteria present.
- They have a very high effectiveness in removing protozoa, bacteria, and viruses.
- They are not effective in removing chemicals.

Water Softeners

- They are used to remove iron and manganese, heavy metals, some radioactivity, nitrates, arsenic, chromium, selenium, and sulfate.
- They use ion exchange technology for chemical or ion removal to lessen the amount of hardness (calcium, magnesium) in the water.
- They offer no defense against bacteria, viruses, or protozoa.

Chapter 6:

Bringing It All Together—The Impact of Water on Our Health

In order to hydrate your epidermis and saturate your brain, muscles, organs, and tissues, how does water actually get to those parts of your body? The answer is fascia. Fascia can be simply defined as tissue. It's a unique, incredibly gauzy, gossamer tissue that sits between and around your organs and bones and beneath your skin. And it stretches for kilometers. We are completely covered in fascia. It is among the most profound anatomical riddles that science is presently delving into.

Just as fascinating now as it was when the first human body was studied is this uncharted region. The easiest way to put it is through the words of American biophysicist Dr. James Oschman, who spent four years studying electrical and fluid transport in cells at Cambridge University: "Fascia forms the largest system in the body, as it is the system that touches all the other systems" (Cohen & Bria, 2018). Furthermore, the role of fascia is crucial to the process of hydration, as we are learning. Unless you have had plantar fasciitis, a frequent but excruciating inflammation of the connective tissue in the arch of your foot, you may not have heard of fascia.

Stretching this fascia restores moisture to the tissue and hastens healing.

However, until recently, fascia was thought of as just a protective covering, similar to Saran Wrap, that held muscles and organs in place. The skeletal, vascular, muscular, and nerve systems, as well as the organs, would be the crucial parts that dissections would remove by peeling and discarding them like plastic wrap. By the time student dissections came to that thrown-out, plastic-like wrap, it was already dehydrated and dried out. It didn't appear to be significant. The first time fascia was observed by scientists in its whole, both alive and hydrated, was in 2005.

That year, a very delicate procedure was carried out by the French surgeon Dr. Jean-Claude Guimberteau, who was very skilled in hand reconstruction. He inserted a fiber-optic camera beneath the patient's epidermis to improve his view of the tissues during the procedure. Usually, crimson blood would have made the intricate fascia underlay difficult to detect, but on this occasion, he clamped off the blood vessels to improve his view of the fascia tissue. Pulsating and undulating, the delicate mesh was captured on video and revealed by the fiber-optic camera, giving the impression that it was breathing. Furthermore, water droplets were being transported by this transparent netting, demonstrating for the first time that fascia is one of the body's main water transport systems.

The irrigation system's expansion and contraction were visible in this video as the water was sent through its tubing. Fascia detection is one of the greatest discoveries of our time, along with gel water. This "aha" moment drastically changed our understanding of how our bodies distribute and store water. The footage made it very evident that the sheets and hollow tubing that make up fascia are what actually let the water you drink enter your tissues.

This finding demonstrates the critical function fascia plays in linking bodily activity and hydration. This water-delivery system is activated by any motion, including twisting, stretching, and rotating. The hydraulic pumping activity of fascia is achieved through constriction and release. It functions exactly like a hydraulic pump. Hydraulics literally translates to "movement by water." This is a genuinely amazing discovery. Rather than being just filler tissue or a protective covering for our body components, fascia is a unique, useful hydraulic system.

Fascia is so essential to the movement of water since it is a clear, nearly undetectable mobile system. Water's journey inside our tissues and cells can now be tracked. In actuality, after drinking, moving is the essential next step in staying hydrated. Water is drawn into our cells through our tissues by movement. It is the other half of being hydrated, quite literally. Hydration is achieved by movement; drinking only initiates it.

Living water is moving water. Think about it. A river or a stream is better than stagnant water. Water is supposed to move through our bodies, and when we activate our fascia, we facilitate this movement. We'll discuss this further in the subchapter on tips for incorporating healthy water habits. We have come full circle and returned to where we started. Water is vital for our health, and drinking high-quality water is crucial. We want our fascia and every part of us to vibrate with vitality.

In this chapter, we provide important tips for incorporating healthy water habits and provide further resources for your enrichment.

Tips for Incorporating Healthy Water Habits

Make wise decisions, such as staying away from sugary, carbonated drinks, and include these suggestions into your daily routine to improve your health and hydration:

Choose a water bottle that corresponds to your intake objectives. Use this container throughout the day after filling it every morning. Once the container is empty, you will have achieved your objective! Make drinking water a "habit." At every meal, before sleeping, and upon waking up, sip water. Be mindful of time. To prevent the day from slipping away from you, try to complete half of your bottle by midday or take a drink every hour. When dining out, go for water to reduce costs and calories. A pitcher of water should always be on the table when dining at home.

Sometimes, we may not necessarily like drinking water because it tastes, well, like water. For flavor, infuse your water with berries, lemon, lime, or cucumber. If you miss the pop of carbonated beverages, try sprinkling some juice over plain sparkling water. Give up purchasing sugar-filled beverages. They can't as easily entice you when they're not at your house. Be mindful of your diet. If something more than just plain water is required, consider the calories, added sugars, and saturated fats.

Movement

The surprising thing is that exercise helps you stay hydrated. Exercise can dehydrate you, yet mobility is essential for maintaining adequate hydration. Numerous studies have demonstrated the critical role that movement and exercise play in maintaining our health. That is not a novel concept. The ground-breaking revelation, however, is that movement has an even more essential role: Hydration is transferred to the cellular level by movement. Additionally, it doesn't require much movement—even the tiniest motion can assist in distributing water throughout your entire body.

A significant study in 2016 involved closely monitoring 12,776 British women with the sole goal of determining whether fidgeting would lengthen life. It did. Women who sat for seven hours or longer a day and showed little to no fidgeting were linked to an astounding 43 percent higher chance of dying from all causes. Contrast that with women who were classified as middle- or even high-fidgeters. Their chance of dying was not higher (Cohen & Bria, 2018).

In actuality, your cells are motion detectors. Your own engine is within you, waiting to be ignited by movement—any movement at all. It is not about running a marathon or spending endless hours at the gym; every region of your body can benefit from the tiniest, most basic motion that helps attract water and electrical charge.

While you move, the pulley-like motion of your skin, muscle, and fascia stretching over your skeletal architecture simultaneously forces your body's cells into action and forces hydration deeper into your tissues. Any movement, no matter how big or small—a big stride, a modest elevation of the shoulders, or a head tilt—stretches the tissues and creates electricity. Piezoelectricity, or "pressure electricity," is the scientific term for this phenomenon, which is based on simple mechanical tension and is a principle found throughout nature.

Hydration percolates into your tissues and charges your cells thanks to piezoelectric action. The bottom line is that the next time you hydrate, perhaps take a walk.

Safe Water Storage

While we have addressed how to treat water, we haven't addressed how to safely store it. The next few chapters are dedicated to this.

When water is stored in a container that keeps it from getting contaminated again, it indicates that the water has been treated and is safe to use. Water can be made safe in various ways. The ideal solution for your family or community will rely on a number of factors, including the quality of the water, the availability of technology, the accessibility of a water supply, the ability to establish hygiene facilities, and other local variables.

When safe water is being carried and stored, contamination can be avoided by using safe storage containers. The ideal container to use while storing clean water is one that is composed of sturdy metal, ceramic, or plastic; features a single, tiny (5–8 cm) hole that is either tightly closed or covered; and has a short neck or hole that allows water to be poured out without getting into the container with hands or things within. Contains a convenient handle for easy carrying and can contain a typical capacity of water (20 liters).

Containers used to store water need to be cleaned frequently. Use this procedure to clean and sterilize a water storage container before adding safe water or treating existing water in it:

- After giving the container a thorough water and soap wash, rinse it out.
- Use a solution consisting of 5 mL (1 tsp) of unscented chlorine bleach in 1 liter of water or 19mL (4 tsp) of unscented chlorine bleach in 1 gallon of water to sanitize the container.
- Use bleach that has a sodium hypochlorite content of 5%–9%.

- Tightly cover the container and give it a good shake. Ensure that the inner surface of the container is touched by the bleach solution.
- Pour the bleach solution out of the container after at least 30 seconds of waiting.
- The empty sterilized container can either be rinsed with safe (treated) water before use or left to air dry for storage.
- Fill the sanitized container with safe or to-be-treated water, then seal it tightly.
- If the water hasn't been treated previously, treat it.
- Put the date of treatment on the container and mark it as "drinking water."

Water should be kept in a cool environment. Water containers should not be kept in direct sunlight or near harmful materials like pesticides or fuel.

Resources for Further Exploration

Public drinking water quality and your local drinking water:

- www.cdc.gov/nceh/tracking/topics/drinkingwater.htm
- www.cdc.gov/nceh/tracking/topics/drinkingwater.htm
- www.epa.gov/ccr/ccr-information-consumers
- www.epa.gov/sdwa/how-epa-regulates-drinking-water-contaminants
- www.epa.gov/safewater

How to protect and improve well water quality:

- www.cdc.gov/healthywater/drinking/groundwater-awareness-week.html
- www.cdc.gov/nceh/ehs/water/private-wells/index.html
- www.cdc.gov/nceh/ehs/elearn/swpi.html
- www.epa.gov/safewater/privatewells/index2.html
- www.privatewellclass.org/

Other drinking water and health-related resources:

- www.cdc.gov/healthywater/disease/az.html
- www.cdc.gov/fluoridation
- www.cdc.gov/nutrition/data-statistics/plain-water-the-healthier-choice.html
- www.cdc.gov/nceh/hsb/cwh/default.htm

History of Drinking Water Treatment

- www.cdc.gov/healthywater/drinking/history.html

Drinking Water

- AAP: Committee on Environmental Health and Committee on Infectious Diseases. pediatrics.aappublications.org/content/123/6/1599 .full.pdf+html Pediatrics. 2009;123(6):1599-1605.

- NIH: Flowers CB, Mackar R. www.niehs.nih.gov/news/newsroom/releases/2009/may26/ National Institute of Environmental Health Sciences. 2009.

- AAP: Rogan WJ, Brady, MT, the Committee on Environmental Health, and the Committee on Infectious Diseases. pediatrics.aappublications.org/content/123/6/e1123.full.pdf+html Pediatrics. 2009;123(6):e1123-37.

- Braff-Guajardo E, Hecht K, Hampton KE, Brindis CD, Grumbach JM, Patel AI. changelabsolutions.org/publications/free-water-k-12 2012.

Waterborne Disease (General)

- CSTE. c.ymcdn.com/sites/www.cste.org/resource/resmgr/PS/09-ID-02.pdf 2009.
- CSTE. c.ymcdn.com/sites/www.cste.org/resource/resmgr/PS/06-ID-12FINAL.pdf 2006.
- CSTE. c.ymcdn.com/sites/www.cste.org/resource/resmgr/PS/06-ID-06FINAL.pdf 2006.
- AAP: Greer FR, Shannon M, the Committee on Nutrition, and the Committee on Environmental Health. pediatrics.aappublications.org/content/116/3/784.long Pediatrics. 2005;116(3):784-6.

Cryptosporidiosis

- CSTE. c.ymcdn.com/sites/www.cste.org/resource/resmgr/PS/11-ID-14.pdf 2010.
- CSTE. c.ymcdn.com/sites/www.cste.org/resource/resmgr/PS/08-ID-08.pdf 2008.
- CSTE. c.ymcdn.com/sites/www.cste.org/resource/resmgr/PS/09-ID-32.pdf 2009.
- CSTE. www.cste2.org/ps/1994/1994-11.htm 1994.
- CSTE. www.cste2.org/ps/1994/1994-nsc-06.htm 1994.

Giardiasis

- CSTE. c.ymcdn.com/sites/www.cste.org/resource/resmgr/PS/10-ID-17.pdf 2010.
- CSTE. c.ymcdn.com/sites/www.cste.org/resource/resmgr/PS/2001-ID-8.pdf 2001.

Conclusion

Tucked away between Dogwood Butte and Thimbleberry Ridge in Northern California, McCloud is a small hamlet with 1,300 residents, one grocery store, one wayside hotel, and no stoplights. Surrounded by forests of fir trees, some of the best trout-fishing streams in the nation, and Mount Shasta towering over everything, it's incredibly picturesque. The town's sole industry for most of the previous century has been logging. The community was essentially the McCloud River Lumber Company, also lovingly referred to as "Mother McCloud." The hotel, the grocery store, and even the homes were owned by the firm.

Ma McCloud was a loving mother. The town received its electricity from the mill, and all it needed for a team from Mother McCloud to come and fix problems like a leaking kitchen faucet or broken lights was a simple phone call. The families of workers received complimentary turkeys on Thanksgiving every year. At the local theater, the curtain was opened for the Christmas celebration to reveal an enormous pile of wrapped presents for every youngster. Life was pleasant. The town's gradual downfall started in the 1960s when the company's fortunes started to shift. Houses were traded in. Local companies failed. Individuals dispersed. When the mill closed in 2002, the former middle-class loggers were left with lowly jobs and food stamps. From a student body of one hundred fifty, the high school suddenly had less than twelve. For years, the school had been unable to have a football team. It appeared to be a cruel joke, with the large lettering spelling out "HOME OF THE

LOGGERS" on the exterior of the gym. McCloud was near death.

A few local government officials in McCloud became desperate to find work and began searching for alternatives to lumber. They must have taken a bold step when they bet the town's future on the town's water supply, another abundant resource in the area. A significant portion of Mount Shasta's glacial melt flows underground and resurfaces lower down the hill. Clear, refreshingly cold, drinkable water gushes from these springs. People in the area have made jokes about McCloud's youthful appearance, claiming that it stems from his fountain of youth. New Age travelers also believe that there is something unique about the water. Some regard this location to be a spiritual vortex, and some travel as far as Japan to meditate there.

Before retiring, Doris Dragseth co-owned the neighborhood auto parts store with her husband. It must have appeared like a strange decision to her and the other district board members of McCloud to replace timber with water. There had always been free spring water. Locals filled a gallon jug as they strolled up. However, it was obvious that things were shifting. It was evident from soft drink vending machines and grocery store shelves that the bottled water industry was seeing significant growth nationwide. More than nine billion gallons of bottled water were consumed in the United States in 2011; that is, around 312 single-serve bottles for every man, woman, and child. And the figures continue to rise. It looks like bottled water may soon overtake soft drinks as the most popular beverage in the country.

Dragseth believed that this was a market that they ought to be in. In the end, McCloud's spring water was probably even superior to anything that was bottled. The McCloud River

Lumber Company had closed its doors, but perhaps the nearby waterways could be exploited by a new Mother McCloud. At the time, the town's future also appeared far-fetched. The district board of McCloud made an effort to persuade water firms to construct a bottling factory in the late 1990s. Until Nestlé Waters North America, a division of the enormous Swiss food and beverage corporation, showed interest, they had no luck. Nestlé, the biggest bottled water business in the United States, held a nearly thirty percent market share with well-known brands like San Pellegrino, Perrier, and Poland Spring. What Nestlé observed pleased it. It's become harder and harder to locate large sources of pure spring water with minimal dissolved minerals, which American consumers desire as a clean, flat flavor. In the Sierra Nevada mountains, for instance, some of California's greatest groundwater is tainted with MTBE, an additive that makes gas burn cleaner but is also carcinogenic to humans.

Due to financial constraints, the McCloud district board was unable to hire legal counsel; instead, Nestlé covered the costs of outside consultants, permits, and legal expenses. The proposed deal was better than supporters of McCloud could have imagined: it included an annual payment to the town, a contract to bottle the town's waters for fifty years with an option for fifty more, a one-million-square-foot bottling plant on the site of the former lumber mill, and the promise of steady jobs. For Dragseth, the decision was clear-cut: "It is beyond me why anyone on the planet would oppose this. We require both the work and the money" (Salzman, 2017).

Approximately one hundred McCloud locals attended a public meeting on September 29, 2004, to learn more about the idea. Nestlé displayed various PowerPoint presentations along with a business film. Each of the five district board members cast a vote in favor of the plan. Not a conversation. Agreement reached. Local company owner Richard McFarland was sitting in the audience, looking surprised. He had never heard of the arrangement before. Not a silent observer of a vote, but one who would discuss the advantages and disadvantages. Over the next three days, he became increasingly enraged as he gathered more details.

This operation was quite large. Nestlé used to extract up to 520 million gallons of water a year from nearby springs, package it under the Arrowhead brand, and load up to 600 semitrailers with bottles every day. According to McFarland, Nestlé was posing a threat to empty the aquifer, destroy the ecosystem of the nearby waterways, and fill McCloud with pollution, noise, and truck traffic. That meant that McCloud was getting paid almost nothing in return—just one penny for every seventeen gallons that Nestlé bottled and subsequently sold for at least forty-five dollars at retail. "The contract read like nobody on the services district knew what they were doing," observed McFarland (Salzman, 2017). It seems as though Nestlé's legal team drafted it.

And McFarland wasn't alone. Gallery owner Janet Connaughton said, "We feel like we've been robbed." This town's birthright shouldn't be sold for a few bucks. Curtis Knight, manager of a conservation organization that protects fisheries, was concerned about how Nestlé's planned extraction of water from aquifers would affect the ecosystem. Imagine if the water levels dropped. He declared, "Water from the McCloud River is sacred. It is regarded globally as one of the most cherished and well-liked trout fishing streams in the nation" (Salzman, 2017). The McCloud Watershed Council was established with the help of these and other allies, with the express purpose of opposing the project. In the courts and the media, it has battled Nestlé.

This tight-knit community was negatively impacted by the well-intentioned attempt to save it, as locals chose sides in a more acrimonious dispute. Opponents of the proposal, many of the longtime residents, were resentful and called newcomers from other parts of California "Flatlanders." Former mill worker Ron Berryman expressed his dissatisfaction, saying, "These people did everything they could to destroy the timber industry." They now want us to turn down a factory that produces bottled water. How much cleaning is possible? Bottled air?

In McCloud, Debra Anderson's family has four generations of history. She was a strong opponent of the Accord and had the tires of her car cut. Four incendiary bombs were discovered in a Nestlé bottling factory pumping station during a comparable fight in Michigan. Although the bombs did not detonate, the Earth Liberation Front took credit for them. The note its members left was a clear warning: "We will no longer stand idly by while corporations profit at the expense of all others. To achieve this, we have intervened at one of the pumping stations that Perrier utilizes to pilfer water. One of the most basic needs is clean water, which no one should be able to commercialize, privatize, or attempt to sell back to us."

The narrative of McCloud revolves around water and a town down on its luck that is sharply split over its desired future—about the kind of future that it requires. However, it goes much beyond that. Stories like these may be written about battles fought overseas in places like Cochabamba, Bolivia; Kerala, India; the West Bank and Gaza Strip; or fights in Fryeburg, Maine; Mecosta County, Michigan; Groveland,

Florida. Conflicts over drinking water are becoming more frequent, not just at the local level but also across the nation and the world.

These disputes may center on a straightforward and plentiful topic, but the essence of the disputes is far from straightforward. There has been much myth and conflict, science and religion, ethics and commerce, in the history of drinking water. The McCloud narrative is derived from a much larger, far older story; therefore, it should not be surprising that it represents these ideas. Human settlement expansion and health have been influenced by our understanding and management of drinking water for over 3,000 years. The situation persists to this day.

Bottled water, which has become "bigger than soft drinks," is something we have explored extensively in this book. We have juxtaposed it with tap water and seen if the claims of bottled water being safer and healthier than tap water are actually substantiated. It isn't. Both bottled water and tap water can be contaminated. When tap water is treated according to the regulations laid out by the EPA, then there is no need for bottled water.

However, while tap water fulfills basic functions, there are waters that are remarkable for our health. DDW and Electrolyzed waters are the living waters that make the tissues in our bodies come alive. These are the champions of water, and there are systems that we can buy to produce these waters. We also discussed other common treatment methods and looked at their limitations with regard to bacteria, viruses, and radiation.

We believe that people have the right to safe drinking water, but as can be seen with the McCloud story above, it is not always black and white. Sometimes, as a citizen, you have to take matters into your own hands and ensure that the water that you provide for yourself and your family is of the highest possible quality. This book's mission was to empower you with the information and tools to be able to do this.

While we have reached the end of this book, the story of water is still being written.

Thanks for purchasing this book.

Please support independent publishers, like me,

by giving your review on amazon

https://mybook.to/essential-water

References

Batmanghelidj, F. (2005). Your body's many cries for water: a revolutionary natural way to prevent illness and restore good health. Tagman.

Bonvissuto, D. (n.d.). What is alkaline water? WebMD. https://www.webmd.com/diet/what-is-alkaline-water

Bottled water vs. tap water: Differences, pros and cons. (n.d.). Svalbarði Polar Iceberg Water. https://svalbardi.com/blogs/water/bottled-vs-tap

Caldecott, J. (2008). Water. Random House.

Importance of water quality and testing. (2020, October 28). Centers for Disease Control and Prevention. https://www.cdc.gov/healthywater/drinking/public/water_quality.html

Cohen, D., & Bria, G. (2018). Quench: beat fatigue, drop weight, and heal your body through the new science of optimum hydration. Hachette Books.

Fleckenstein, A., & Weisman, R. (2007). Health 2.0: Tap into the healing powers of water to fight disease. McGraw Hill Professional.

Fujiiryoki Purifier. (2023, July 25). What is the specialty associated with electrolyzed reduced water? Medium. https://medium.com/@fujiiryokiipurifier/what-is-the-specialty-associated-with-electrolyzed-reduced-water-ec55182546bb

Govender, K. (n.d.). What is deuterium depleted water? Nourishme Organics.

https://www.nourishmeorganics.com.au/blogs/nmo-gut-health-journal/what-is-deuterium-depleted-water

Herzog, S. (2011). Poisoned Golf: A groundskeeper speaks out on contamination and poisoning. Pesticides and You. https://www.beyondpesticides.org/assets/media/documents/infoservices/pesticidesandyou/documents/poisoned-golf2.pdf

Kangen Water - life-changing power of Kangen Water®, https://www.kangenwaternation.com

The inside workings of your body's detox explained. (n.d.). AcuMedic. https://clinic.acumedic.com/detox-a-view-from-the-inside/

Mccauley, R. F. (2006). The miraculous properties of ionized water: the definitive guide to the world's healthiest substance. Se, Inc.

Morris, R. D. (2007). The blue death: disease, disaster and the water we drink. Harpercollins Publishers.

Salzman, J. (2017). Drinking water. Abrams.

7 benefits of water ionizers. (n.d.). Energise Your Life. https://www.energiseyourlife.com/blog/7-benefits-of-water-ionizers/

Villa-Forte, A. (2022, April). Organ systems. https://www.msdmanuals.com/home/fundamentals/the-human-body/organ-systems

What equipment do you need to purify water at home? A comprehensive guide. (n.d.). Mermaid Water Systems. https://mermaidwatersystems.com/what-equipment-o-you-need-to-purify-water-at-home-a-comprehensive-guide/

Made in United States
North Haven, CT
10 November 2026

22088258R00085